Plant Conservation Science and Pra
The Role of Botanic Gardens

Only a green world, rich in plants, can sustain us and the millions of other species with which we share this planet. But, in an era of global change, nature is on the retreat. Like the communities they form, many plant species are becoming rarer, threatened even to the point of extinction. The worldwide community of almost 3,000 botanic gardens are holders of the most diverse living collections of plants and have the unique potential to conserve plant diversity. Conservation biology is a fast moving and often controversial field and, as the contributions within these pages from experts in the field demonstrate, plant conservation is multifaceted, mirroring the complexity of the biodiversity it aims to protect, and striving not just to protect threatened plants but to preserve ecosystem services and secure the integrity of the biosphere.

STEPHEN BLACKMORE is Queen's Botanist and Honorary Fellow of Royal Botanic Garden Edinburgh (RBGE), and Chairman of Botanic Gardens Conservation International (BGCI) and the UK Government's Darwin Initiative. He has formerly been Regius Keeper of RBGE (1999 to 2013) and Keeper of Botany at the Natural History Museum (1990 to 1999).

Awarded an OBE in 2016 for conservation and protection of wild tree species, SARA OLDFIELD is co-chair of the IUCN/SSC Global Tree Specialist Group. From 2005 to 2015, she led the work of Botanic Gardens Conservation International (BGCI) as Secretary General.

ECOLOGY, BIODIVERSITY AND CONSERVATION

Series Editors
Michael Usher   *University of Stirling, and formerly Scottish Natural Heritage*
Denis Saunders   *Formerly CSIRO Division of Sustainable Ecosystems, Canberra*
Robert Peet   *University of North Carolina, Chapel Hill*
Andrew Dobson   *Princeton University*

Editorial Board
Paul Adam   *University of New South Wales, Australia*
H. J. B. Birks   *University of Bergen, Norway*
Lena Gustafsson   *Swedish University of Agricultural Science*
Jeff McNeely   *International Union for the Conservation of Nature*
R. T. Paine   *University of Washington*
David Richardson   *University of Stellenbosch*
Jeremy Wilson   *Royal Society for the Protection of Birds*

The world's biological diversity faces unprecedented threats. The urgent challenge facing the concerned biologist is to understand ecological processes well enough to maintain their functioning in the face of the pressures resulting from human population growth. Those concerned with the conservation of biodiversity and with restoration also need to be acquainted with the political, social, historical, economic and legal frameworks within which ecological and conservation practice must be developed. The new Ecology, Biodiversity, and Conservation series will present balanced, comprehensive, up-to-date, and critical reviews of selected topics within the sciences of ecology and conservation biology, both botanical and zoological, and both 'pure' and 'applied'. It is aimed at advanced final-year undergraduates, graduate students, researchers, and university teachers, as well as ecologists and conservationists in industry, government and the voluntary sectors. The series encompasses a wide range of approaches and scales (spatial, temporal and taxonomic), including quantitative, theoretical, population, community, ecosystem, landscape, historical, experimental, behavioural and evolutionary studies. The emphasis is on science related to the real world of plants and animals rather than on purely theoretical abstractions and mathematical models. Books in this series will, wherever possible, consider issues from a broad perspective. Some books will challenge existing paradigms and present new ecological concepts, empirical or theoretical models, and testable hypotheses. Other books will explore new approaches and present syntheses on topics of ecological importance.

Ecology and Control of Introduced Plants
Judith H. Myers and Dawn Bazely

Invertebrate Conservation and Agricultural Ecosystems
T. R. New

Risks and Decisions for Conservation and Environmental Management
Mark Burgman

Ecology of Populations
Esa Ranta, Per Lundberg and Veijo Kaitala

Nonequilibrium Ecology
Klaus Rohde

The Ecology of Phytoplankton
C. S. Reynolds

Systematic Conservation Planning
Chris Margules and Sahotra Sarkar

Large-Scale Landscape Experiments: Lessons from Tumut
David B. Lindenmayer

Assessing the Conservation Value of Freshwaters: An International Perspective
Philip J. Boon and Catherine M. Pringle

Insect Species Conservation
T. R. New

Bird Conservation and Agriculture
Jeremy D. Wilson, Andrew D. Evans and Philip V. Grice

Cave Biology: Life in Darkness
Aldemaro Romero

Biodiversity in Environmental Assessment: Enhancing Ecosystem Services for Human Well-being
Roel Slootweg, Asha Rajvanshi, Vinod B. Mathur and Arend Kolhoff

Mapping Species Distributions: Spatial Inference and Prediction
Janet Franklin

Decline and Recovery of the Island Fox: A Case Study for Population Recovery
Timothy J. Coonan, Catherin A. Schwemm, and David K. Garcelon

Ecosystem Functioning
Kurt Jax

Spatio-Temporal Heterogeneity: Concepts and Analyses
Pierre R. L. Dutilleul

Parasites in Ecological Communities: From Interactions to Ecosystems
Melanie J. Hatcher and Alison M. Dunn

Zoo Conservation Biology
John E. Fa, Stephan M. Funk and Donnamarie O'Connell

Marine Protected Areas: A Multidisciplinary Approach
Joachim Claudet

Biodiversity in Dead Wood
Jogeir N. Stokland, Juha Siitonen, and Bengt Gunnar Jonsson

Landslide Ecology
Lawrence R. Walker and Aaron B. Shiels

Nature's Wealth: The Economics of Ecosystem Services and Poverty
Pieter J. H. van Beukering, Elissaios Papyrakis, Jetske Bouma and Roy Brouwer

Birds and Climate Change: Impacts and Conservation Responses
James W. Pearce-Higgins and Rhys E. Green

Marine Ecosystems: Human Impacts on Biodiversity, Functioning and Services
Tasman P. Crowe and Christopher L. J. Frid

Wood Ant Ecology and Conservation
Jenni A. Stockan and Elva J. H. Robinson

Detecting and Responding to Alien Plant Incursions
John R. Wilson, F. Dane Panetta and Cory Lindgren

Conserving Africa's Mega-Diversity in the Anthropocene: The Hluhluwe-iMfolozi Park Story
Joris P. G. M. Cromsigt, Sally Archibald and Norman Owen-Smith

Plant Conservation Science and Practice

The Role of Botanic Gardens

Edited by
STEPHEN BLACKMORE
Botanic Gardens Conservation International (BGCI), Richmond, Surrey, United Kingdom

SARA OLDFIELD
Cambridge, United Kingdom

CAMBRIDGE
UNIVERSITY PRESS

University Printing House, Cambridge CB2 8BS, United Kingdom

One Liberty Plaza, 20th Floor, New York, NY 10006, USA

477 Williamstown Road, Port Melbourne, VIC 3207, Australia

4843/24, 2nd Floor, Ansari Road, Daryaganj, Delhi – 110002, India

79 Anson Road, #06–04/06, Singapore 079906

Cambridge University Press is part of the University of Cambridge.

It furthers the University's mission by disseminating knowledge in the pursuit of education, learning and research at the highest international levels of excellence.

www.cambridge.org
Information on this title: www.cambridge.org/9781107148147
DOI: 10.1017/9781316556726

© Cambridge University Press 2017

This publication is in copyright. Subject to statutory exception
and to the provisions of relevant collective licensing agreements,
no reproduction of any part may take place without the written
permission of Cambridge University Press.

First published 2017

Printed in the United Kingdom by TJ International Ltd., Padstow, Cornwall

A catalogue record for this publication is available from the British Library

Library of Congress Cataloging-in-Publication Data
Names: Blackmore, Stephen, editor. | Oldfield, Sara, editor.
Title: Plant conservation science and practice : the role of botanic gardens / edited by Stephen Blackmore, Sara Oldfield.
Other titles: Ecology, biodiversity, and conservation.
Description: New York, NY : Cambridge University Press, 2017. | Series: Ecology, biodiversity and conservation | Includes bibliographical references and index.
Identifiers: LCCN 2017027615 | ISBN 9781107148147 (alk. paper)
Subjects: LCSH: Plant diversity conservation. | Conservation biology. | Botanical gardens.
Classification: LCC QK86.A1 P565 2017 | DDC 333.95/316–dc23
LC record available at https://lccn.loc.gov/2017027615

ISBN 978-1-107-14814-7 Hardback
ISBN 978-1-316-60246-1 Paperback

Cambridge University Press has no responsibility for the persistence or accuracy of URLs for external or third-party internet websites referred to in this publication, and does not guarantee that any content on such websites is, or will remain, accurate or appropriate.

Contents

List of Contributors	*page* xi
Foreword	xiii
Ghillean T. Prance	
Preface	xv
Acknowledgements	xviii

1 Mounting a Fundamental Defence of the Plant Kingdom   1
Stephen Blackmore and Sara Oldfield
    1.1 Introduction   1
    1.2 The Anthropocene   1
    1.3 Botanic Gardens and Global Change   8

2 Using DNA Sequence Data to Enhance Understanding and Conservation of Plant Diversity at the Species Level   23
Peter M. Hollingsworth, Linda Neaves and Alex D. Twyford
    2.1 The Use of Genetics to Tell Species Apart   23
    2.2 Sequencing a Few Loci to Obtain Broad-Brush Patterns: DNA Barcoding   29
    2.3 High Resolution Studies on Complex Groups (Sequences from Multiple Nuclear Loci)   35
    2.4 Future Prospects for Large-Scale Use of Sequence Data at the Species Level in Plants   40
    2.5 Concluding Remarks   41

3 Conservation Assessments and Understanding the Impacts of Threats on Plant Diversity   49
Malin Rivers
    3.1 Habitat Loss   51
    3.2 Overexploitation   51

	3.3	Invasive Species, Pests and Diseases	54
	3.4	Climate Change	56
	3.5	Conservation Assessments	58
	3.6	The IUCN Red List Categories and Criteria	60
	3.7	Influencing Policy and Legislation	68
	3.8	Conclusion	69

4 The Role of Botanic Gardens in *In Situ* Conservation — 73
Jin Chen, Richard T. Corlett and Charles H. Cannon

4.1	Introduction	73
4.2	An Expanded Concept of *In Situ* Conservation	74
4.3	What Can Botanic Gardens Offer to *In Situ* Conservation?	78
4.4	How Are Botanic Gardens Best Involved in *In Situ* Conservation?	79
4.5	Surveys and Inventories	82
4.6	Monitoring of Biodiversity *In Situ*	85
4.7	Ecological Restoration of Degraded Areas	88
4.8	Assisted Migration in Response to Climatic Change	89
4.9	Ethnobotany and the Sustainable Use of Natural Resources	89
4.10	Set up Partnerships with all Stakeholders	90
4.11	Expanding Impact by Networking	94
4.12	Challenges and Opportunities	95
4.13	Conclusions	97

5 The Role of Botanic Gardens in *Ex Situ* Conservation — 102
Paul Smith and Valerie Pence

5.1	Introduction: The Range of Plant Diversity Held by Botanic Gardens in their Living Collections	102
5.2	Living Collections	104
5.3	Seed Banks	106
5.4	Exceptional Species	110
5.5	The Strengths and Weaknesses of *Ex Situ* Collections for Conservation Purposes	120
5.6	Challenges Associated with Maximising the Value of *Ex Situ* Collections for Conservation	122
5.7	The Conservation Opportunities Afforded by *Ex Situ* Collections	126

6	**The Role of Botanic Gardens and Arboreta in Restoring Plants**	
	From Populations to Ecosystems	134
	Kayri Havens	
	6.1 Introduction	134
	6.2 Types of Restoration Work Undertaken by Botanic Gardens	135
	6.3 Overcoming Challenges	149
	6.4 Conclusions	158
7	**Botanic Gardens and Solutions to Global Challenges**	166
	Samuel F. Brockington and Beverley J. Glover	
	7.1 Introduction	166
	7.2 Global Food Security	166
	7.3 The Role of Botanic Gardens in Food Security	168
	7.4 Global Fuel Security	177
	7.5 Education and Public Awareness of Global Challenges	185
	7.6 Conclusions	187
8	**Cultivating the Power of Plants to Sustain and Enrich Life**	
	How Public Gardens Can Realise our Purpose by Focusing on the Basic Human Needs Universal to Diverse Audiences	192
	Sophia Shaw and Jennifer Schwarz Ballard	
	8.1 Reframing 'Community Outreach' as Authentic Engagement	194
	8.2 How Can Botanic Gardens Broaden their Reach and Make a Difference?	196
	8.3 Conclusion: Change Won't Come Easily, but the Stakes are High and the Benefits Are Well Worth the Effort	199
9	**Botanic Gardens and Conservation Impact**	
	Options for Evaluation	219
	Sara Oldfield and Valerie Kapos	
	9.1 Botanic Gardens: How They Operate and How They Are Perceived	219
	9.2 Measuring the Performance of Botanic Gardens	223

		9.3 Measuring Conservation Success	226
		9.4 Measuring Botanic Garden Progress Towards the Targets of the GSPC	230
10	**Conclusions**		236
	Stephen Blackmore, Sara Oldfield and Paul Smith		
		10.1 Key Themes from Chapters	238
		10.2 Botanic Gardens and the Future of Plant Conservation	242

Index 249

Colour plates to appear between pp. 142 and 143

Contributors

STEPHEN BLACKMORE Queen's Botanist and Honorary Fellow, Royal Botanic Garden Edinburgh and Chairman of Botanic Gardens Conservation International (BGCI)

SAMUEL F. BROCKINGTON Cambridge University Botanic Garden, Cambridge, UK

CHARLES H. CANNON Center for Tree Science, The Morton Arboretum, Lisle, Illinois, USA

JIN CHEN Xishuangbanna Tropical Botanical Garden, Chinese Academy of Sciences, Mengla, Yunnan, China

RICHARD T. CORLETT Xishuangbanna Tropical Botanical Garden, Chinese Academy of Sciences, Mengla, Yunnan, China

BEVERLEY J. GLOVER Cambridge University Botanic Garden, Cambridge, UK

PETER M. HOLLINGSWORTH Royal Botanic Garden Edinburgh, Edinburgh, UK

KAYRI HAVENS Medard and Elizabeth Welch Director, Division of Plant Science and Conservation; Senior Scientist, Chicago Botanic Garden, Chicago, USA

VALERIE KAPOS UN Environment World Conservation Monitoring Centre, Cambridge, UK

LINDA E. NEAVES Royal Botanic Garden Edinburgh, Edinburgh, UK; and Australian Centre for Wildlife Genomics, Australian Museum Research Institute, Sydney, NSW, Australia

SARA OLDFIELD Biodiversity consultant and former Secretary General of Botanic Gardens Conservation International (BGCI), Cambridge, UK

VALERIE PENCE Director of Plant Research, Center for Research in Endangered Wildlife (CREW) at Cincinnati Zoo and Botanical Garden, Cincinnati, USA

MALIN RIVERS Botanic Gardens Conservation International (BGCI), Richmond, UK

JENNIFER SCHWARZ BALLARD Vice President, Education and Community Programs, Chicago Botanic Garden, Chicago, USA

SOPHIA SHAW Adjunct Professor, Kellog School of Management, Managing Partner, Acorn Advisors LLC and former President and CEO, Chicago Botanic Garden, Chicago, USA

PAUL SMITH Secretary General, Botanic Gardens Conservation International (BGCI), Richmond, UK

ALEX D. TWYFORD Ashworth Laboratories, Institute of Evolutionary Biology, The University of Edinburgh, Edinburgh, UK

Foreword

Recent reports show that in spite of all the conventions and legislation, the status of the world's plant species continues to decline as was demonstrated in the recent document on the *State of the World's Plants* (RBG Kew, 2016). The *Global Biodiversity Outlook 4* (Secretariat of the Convention on Biological Diversity 2014) paints a disappointing picture of the efforts of governments to meet the set of conservation targets that were agreed in 2002 in the *Global Strategy for Plant Conservation* (Secretariat of the Convention on Biological Diversity 2011) of the Convention on Biological Diversity. It is a report about slow progress in meeting the conservation targets set for 2020. Habitat destruction, the introduction of alien species and climate change are all taking their toll on plant species. I write this not just from reading the reports of others, but as a botanist who has witnessed the deteriorating situation in many parts of the world over my 50 years of plant exploration. The burning forests of Amazonia, the alien plants ousting the native vegetation of Hawaii and acres of oil palms in Borneo have all been part of my experience in the field. We are fast diminishing our options for the future as plants are basic to life on Earth and diversity is essential to maintain their many ecological roles as well as their many uses to human society. The extinction rate is increasing especially on tropical islands such as Hawaii and Madagascar and in the other examples given by the authors of this book. This means that conservation of plant species has of necessity become a major focus of the progammes of most botanic gardens around the word who hold both many living collections and a mass of scientific data about plants. It has also unfortunately meant that *ex situ* conservation has become ever more important, but as this book clearly demonstrates, botanic gardens have taken up the challenge of working on conservation on site as well as off site in seed banks and in their living collections. The coordination of conservation efforts for plants has become ever more important, so the roles of Botanic Gardens Conservation International, the Center for Plant Conservation in the USA and many other national and international organisations are vital for our future.

Here we have a collection of chapters that brings plant conservation science right up to date from modern laboratory techniques such as DNA bar-coding to the latest fieldwork in sensitive regions. My hope is that the contents of this important volume will not just reach those of us who practising conservation, but also the funders and the politicians who are often ready to make rules but slow to follow them up. I read this book as a call for increased and united action to save the plant species of the world by the botanic gardens of the world.

Ghillean Prance FRS, VMH

References

RBG Kew (2016). The State of the World's Plants Report 2016. Royal Botanic Gardens, Kew, available online at https://stateoftheworldsplants.com/areas-important-for-plants [accessed February 2017].

Secretariat of the Convention on Biological Diversity (2014). Global Biodiversity Outlook 4. Montréal: CBD, 155 pp.

Secretariat of the Convention on Biological Diversity (2011). COP 10 Decision X/17. Consolidated update of the Global Strategy for Plant Conservation 2011–2020, available online at www.cbd.int/decision/cop/?id=12283 [accessed February 2017].

Preface

We were delighted to be invited to contribute to the prestigious Ecology, Biodiversity and Conservation series published by Cambridge University Press because, when compared to the conservation of animals, too little attention is directed to the conservation of plants. As Kay Havens has emphasised, citing the example of federal funding in the US, spending on threatened and endangered animals is thousands of times greater than on plants. More positively, however, our experiences of working with botanic gardens have persuaded us of the incredible importance and potential contribution of the global community of some 2670 botanic gardens.

However, we approached the task with considerable trepidation. Conservation biology is a fast moving and often controversial field. A wide diversity of opinions exist even over the fundamental definitions and goals of conservation. At its narrowest, plant conservation is often conceived as focusing on the survival of a single endangered species or the designation and protection of a particular area important for its plants. At the other end of the spectrum, conservation is seen as the actions needed to preserve ecosystem services, or evolutionary potential, or the wider efforts needed to secure a healthy biosphere in an era of global change. The extent of these controversies and the passion with which the protagonists present their case is demonstrated by several continuing debates.

These have seen, for example, a traditional focus on conservation in protected areas set against counter arguments that too little of the natural world, if any of it actually remains, is represented within protected areas. The case is also made that protected areas cannot be isolated from or protected against climate change, especially in an era of global change. The definition and relevance to conservation of novel ecosystems has been hotly debated with some arguing that even accepting that conservation can be conducted in novel landscapes provides developers with a licence to destroy what little remains of pristine ecosystems and their biodiversity. Efforts to repair the damage done, especially since the Industrial Revolution, are equally contentious. There has been intense discussion of the scope, aims and methodology of ecological restoration and the validity of rewilding. There are those who argue for the necessity of assisted migration of threatened species and those who consider it heavy-handed interference. For many years the idea of *ex situ* conservation of threatened plants in botanic gardens, arboreta or seed banks was seen as an admission of failure. Only continuing survival *in situ* could be considered a true conservation success. There are those who still hold to this view.

Strongly divergent opinions are also held on who the principal players are in plant conservation. Is it those who manage protected areas whether they be private individuals, agencies of government or in international programmes such as UNESCO's Man and the Biosphere Programme? Is it the professional forestry community who manage many of the world's forests or those conservation agencies, mining corporations and others who focus on ecological restoration? What is the role of the scientific community in conservation, as theoretical researchers or hands-on practitioners? Many have questioned whether botanic gardens, the focus of this book, even have a significant role to play in conservation beyond the cultivation or seed banking of rare species *ex situ*.

We reject the polarised perspectives that are so often presented on such issues. They follow a time-honoured form of academic discourse, but in nature one size does not fit all. Conservation is a highly complex field, mirroring the complexity of the biodiversity it aims to protect. Consequently, very different approaches are required in different contexts.

There are undoubtedly roles for many players, the more the merrier, given that present efforts are failing to deliver on the internationally agreed agendas of the multilateral environmental agreements. It is equally certain that while protected areas will continue to have enormous importance as the 'best' pieces of our planet, novel ecosystems and agricultural or urban landscapes represent the only future home for many species and must also be part of the mix. Future generations will judge us for the biodiversity we protect and bequeath them, not for the quality of our academic debates in a time of biodiversity crisis and unprecedented environmental change.

Botanic gardens have a unique role to play as centres of expertise in plant diversity and holders of living collections of much greater diversity than are held by any other professional organisations concerned with plants. However, they alone are not sufficient to face the challenges of the Anthropocene. Botanic gardens must continue to diversify the range of partners they work with in order to maximise their contribution to the biodiversity crisis and they must step up the scale and intensity of their efforts to conserve plants. There have been many plant extinctions since humanity encircled the globe and already there are dozens, perhaps hundreds, of species, extinct in the wild, which survive only in cultivation within living collections or seed banks.

We hope, therefore, that this book will show not only what is already being accomplished by botanic gardens in the field of conservation but also what might be possible if and when botanic gardens join forces with the widest possible community of interest, to work together for a green planet. Only a, literally, green world can sustain us and the millions of other species with which we share this planet for the future.

Acknowledgements

We would like to thank Michael Usher for encouraging us to contribute this book to the Ecology, Biodiversity and Conservation series of Cambridge University Press. Our views and experiences of plant conservation have been shaped by our extensive travels to many places where nature is on the retreat, to places where efforts are underway to reverse the damage done in the past and to botanic gardens large and small but equal in their passion and commitment. We have been fortunate to meet many of the leading figures in biodiversity conservation and to benefit from discussions with a host of experts who have shared their ideas and wisdom freely. From those many we were able to select a small number of talented people to contribute to the book and we thank them all for their contributions.

Our thanks are due to Paul Smith, Suzanne Sharrock and other staff at Botanic Gardens Conservation International who have provided reference materials and suggestions for various chapters as well as providing ongoing information services on the work of botanic gardens around the world. We acknowledge the support from Caroline Pollock of the IUCN Global Species Programme in preparing statistics on the IUCN Red List used in Chapter 2 and IUCN for use of Figure 3.2. We are grateful to the team at Cambridge University Press for managing the publication process so smoothly and effectively.

1 · *Mounting a Fundamental Defence of the Plant Kingdom*

STEPHEN BLACKMORE AND SARA OLDFIELD

1.1 Introduction

It is a sobering thought that, during our lifetimes, the state of the global environment has changed more rapidly than at any previous period in human history. Against this background, our particular focus is on the conservation of plant diversity, a fundamental component of the biology of our planet. The purpose of this book is to show, through the contributions of leading lights in the field, how action is being taken to address the biodiversity crisis, with botanic gardens progressively developing and implementing targeted responses. Taken together the efforts of botanic gardens around the world can and should amount to a fundamental defence of the plant kingdom.

This chapter sets the scene by first reviewing the broad background of environmental change during the Anthropocene, exploring the origins of awareness of human impacts on the biosphere and discussing the development of coordinated international responses to these global challenges. We then explore the special contribution of botanic gardens, arboreta and seed banks to the conservation of plant biodiversity and summarise some of the more contentious issues before outlining the scope of the chapters that follow.

1.2 The Anthropocene

The pace of change in the global environment began to accelerate markedly as the human population passed one billion in the early nineteenth century with the start of the Industrial Revolution (Steffen *et al.*, 2007; Blackmore, 2009). However, as Ellis (2011) and others have emphasised, the modification of nature by humans extends back long into prehistory, well before the development of settled agriculture. Whereas for millennia the pace of change was slow and gradual, the collective impact of humanity on the biosphere is now so great that we are living through the early years of the Anthropocene, a new period in the geological history of the planet (Crutzen 2002; Steffen *et al.*, 2007; Steffen *et al.*, 2011). Humankind has become a global geological force in its own right, with profound implications for the future.

In recent years, numerous reviews (see, for example, Mooney, 2010; Ellis, 2011; Brook *et al.*, 2012; Ellis *et al.*, 2013) have documented the extent to which the burgeoning demands of humanity outstrip the capacity of nature to replenish and renew and are, therefore, by definition, unsustainable. One powerful and widely discussed way of representing this is the framework established by Rockström *et al.* (2009) based on a set of nine planetary boundaries intended to delimit a safe operating environment for humanity. When this important paper was first published, three of the boundaries it proposed had already been crossed, with the loss of biodiversity being the most over-stretched of all. An update by the same team just six years later concluded that four planetary boundaries have now been crossed (Steffen *et al.*, 2015). Defining a planetary boundary for biodiversity has proved to be difficult. Mace *et al.* (2014) pointed out that the original biodiversity planet boundary was a relatively unsophisticated measure, focusing narrowly on species extinction without adequately measuring the full impact of biodiversity loss on the needs of humanity. These authors suggested three facets of biodiversity on which a redefined boundary might be based: the genetic library of life, functional type diversity and biome condition and extent. They also suggested that the importance of biodiversity in providing a safe operating space for humanity may lie primarily in its interactions with the other boundaries. This perspective is not intended to reduce the inherent importance of biodiversity or to imply that it is any less critical than any of the other boundary indicators. We strongly endorse the importance of recognising the complex interconnections in the Earth system and the limitations of considering issues such as climate change, food security, biodiversity loss, poverty alleviation, desertification or other dimensions of global change in isolation. Unfortunately these issues are, too often, considered individually or ranked in relative importance as though they were not intimately interlinked. At the same time we recognise the complexity of the issues and the, as yet, incomplete knowledge and understanding of biological systems within which we operate. Within the wider field of biodiversity conservation which is an essential component of action to address food security and the other major issues referred to above, the purpose of this book is to consider the best ways of understanding and maintaining plant biodiversity as a fundamental precondition for a sustainable future.

Although awareness of global change and the impact of humanity on nature are not, in reality, recent phenomena (see Steffen *et al.*, 2011, for a review), it is only in recent decades that these issues have begun to shape and define the international policy agenda. As early as 1864, George P. Marsh wrote that '... the action of man upon the organic world tends to subvert the balance of its species, and while it reduces the numbers of some of them, or even extirpates them altogether, it multiplies other forms of animal and vegetable life' (Marsh, 1864, p. iv). His far-sighted book concluded with 'a chapter upon Probable and Possible Geographical Revolutions yet to be effected by the art of man'.

Awareness of the extent and impact of the demands we make on the planet grew incrementally through the twentieth century. At first this awareness was fed by individual expressions of the need to celebrate and protect the natural environment generally for its scenic magnificence and iconic species (see, for example, Muir 1911, 1992), and later by calls for a new ethical basis to our relationship with nature (Leopold, 1968). Following the Second World War, moves to form an international environmental organisation initially focused on the conservation of wildlife rather than on the use of natural resources (Adams, 2004). The International Union for the Protection of Nature (IUPN) was formed in 1947 and changed its name to the International Union for Conservation of Nature and Natural Resources (IUCN) in 1956, reflecting a broader remit. As one component of its broad work, the IUPN established a 'Survival Service' that took on responsibility for cataloguing threatened species of plants and animals and in 1956 this became a permanent Commission of IUCN. The first Red Data book, a concept devised by the hugely influential conservationist Sir Peter Scott, was published in 1962.

From the 1960s onwards, concerns about the unintended consequences of our actions (for example, Carson, 1962) led to the growth of an environmental movement in civil society. The World Wildlife Fund (now World Wide Fund for Nature, WWF) was launched in 1961, initially to fund the work of IUCN, and Friends of the Earth was launched ten years later. More recently, Lovelock (1988, 2006), Wilson (1992, 2002) and others have emphasised the complex, interconnected nature of the Earth's living systems, and the need to preserve this complexity, not merely the elements it is made up of. This fundamental characteristic of the web of life was elegantly expressed in John Muir's often quoted (and misquoted) lines, 'When we try to pick out anything by itself, we find it hitched to everything else in the Universe' (Muir, 1911).

The growing awareness created by various strong and influential voices led to environmental campaigning which, in turn, progressed into concerted international efforts at a political level. In June 1972 the first UN summit on the environment, the United Nations Conference on the Human Environment, was held in Stockholm, placing environmental issues on the global political agenda. The need to increase awareness of the economic, social and political effects of environmental problems was recognised. One outcome of the Conference was the formation of the United Nations Environment Programme (UNEP). The initial objectives of UNEP were broad including to improve knowledge of ecological systems, to improve planning for development that took into account environmental considerations and to build capacity for preservation and enhancement of the environment.

During the 1970s efforts continued at national, regional and international levels to catalogue the conservation status of plants and animals, as one strand of environmental action, and to use this information to inform legal protection. In the US, for example, the Endangered Species Act came into force in 1973 with listing of endangered and threatened species subject to legal protection. Internationally the Convention on International Trade in Endangered Species of Fauna and Flora (CITES) came into force in 1976 with appendices listing species at risk through overexploitation for international trade.

By the time of the second UN environmental summit held in Rio de Janeiro in 1992, there was global recognition of the inextricable links between environment and development expressed, for example, by the Brundtland Report (United Nations, 1987). One important outcome of the Rio meeting was the Convention on Biological Diversity (CBD) with its three objectives: conserving biodiversity, ensuring the sustainable use of biodiversity and ensuring the fair and equitable sharing of benefits arising from the use of genetic resources. Another equally important outcome was the United Nations Framework Convention on Climate Change (UNFCCC). Under the auspices of the United Nations and other international agencies, global biodiversity conservation efforts have continued to develop through the suite of Multilateral Environmental Agreements (MEAs) including the CBD and CITES (both administered by UNEP) that directly embrace the conservation of plant species. The recently established Intergovernmental Science-Policy Panel on Biodiversity and

Ecosystem Services (IPBES) is further evidence of the growing connections between biodiversity research and decision makers (Heywood, 2015).

> Box 1.1 *The International Treaty on Plant Genetic Resources for Food and Agriculture*
>
> The International Treaty on Plant Genetic Resources for Food and Agriculture (ITPGRFA) was adopted by the thirty-first session of the Conference of the Food and Agriculture Organization of the United Nations on 3 November 2001. The Treaty aims to:
>
> - recognise the enormous contribution of farmers to the diversity of crops that feed the world;
> - establish a Global System to provide farmers, plant breeders and scientists with access to plant genetic materials;
> - ensure that recipients share benefits they derive from the use of these genetic materials with the countries where they have originated.
>
> The Treaty helps maximise the use and breeding of all crops and promotes development and maintenance of diverse farming systems. It addresses Target 9 of the GSPC and Aichi Target 13 on the maintenance of genetic diversity. The Treaty's solution to access and benefit sharing – the Multilateral System (MLS) – complies with the Nagoya Protocol, also under the CBD (see Chapter 5). The Treaty puts 64 of the world's most important crops – crops that together account for 80 per cent of the food we derive from plants – into an easily accessible global pool of genetic resources. This is freely available to potential users in the Treaty's ratifying nations for research, breeding and training. The 64 priority crops are listed in Annex 1 of the Treaty.
>
> The Treaty prevents the recipients of genetic resources from claiming intellectual property rights over those resources in the form in which they received them, and ensures that access to genetic resources already protected by international property rights is consistent with international and national laws. Those who access genetic materials through the Multilateral System agree to share any benefits from their use through four benefit-sharing mechanisms established by the Treaty.
>
> Some 128 countries are now contracting parties to the Treaty. As such they commit to 'cooperate to promote the development of

> an efficient and sustainable system of *ex situ* conservation' and require that all parties cooperate to promote the conservation, evaluation and documentation of these resources within a new Multilateral System for access and benefit sharing.
>
> Under Article 14, the Treaty calls on all contracting parties to implement the Global Plan of Action (GPA). The UN Food and Agriculture Organization (FAO) launched the first GPA in 1996 which was adopted by 150 countries. The GPA called for 'safeguarding as much existing unique and valuable diversity as possible in *ex situ* collections of plant genetic resources for food and agriculture' and to 'develop an efficient goal-oriented, economically efficient and sustainable system of *ex situ* conservation'. The GPA further called on stakeholders to 'develop and strengthen cooperation among national programmes and international institutions to sustain *ex situ* collections'. The GPA has recently been revised, but still maintains this focus.[1]

The fact that these and other MEAs have been adopted by nearly all the nations of the world is, in itself, a remarkable and unprecedented level of international consensus. The MEAs have spawned innumerable small successes and some more significant advances, but they have not yet halted the steady and continuing erosion of Earth's life-support systems. It is fair to say that this failure has not been because the challenges themselves are intractable but rather that too little has been invested in tackling them. Nature has tended to be perceived as a matter of secondary importance, less important than the state of the national and international economy or other seemingly pressing matters of the day. We have yet to take to heart the wisdom of the Brundtland Report, which stated,

The environment does not exist as a sphere separate from human actions, ambitions, and needs, and attempts to defend it in isolation from human concerns have given the very word 'environment' a connotation of naivety in some political circles ... But the 'environment' is where we all live; and 'development' is what we all do in attempting to improve our lot within that abode. The two are inseparable. *(UN, 1987)*

It is important to note that many of the successes that have been achieved around the world reflect the efforts of voluntary and charitable organisations. This provides an indication of the importance that wider society places on the quality of life in a sustainable future, even when short-term political priorities clearly lie elsewhere.

We can only hope that this state of affairs will end soon. It is increasingly evident that such issues as climate change, food security, desertification, biodiversity loss, poverty and human migration all have their roots in the degradation of the natural environment. However, a wider understanding of the interconnected complexities of life on Earth will not, on its own, lead to a sustainable future. Achieving what the Rio+20 conference called, 'The Future We Want' (UNDP, 2012) will require understanding to be followed by determined international leadership and the urgent commitment of significantly greater resources and effort. Perhaps we are finally travelling in the right direction, albeit far too slowly. Steffen et al. (2007) described the emergence of efforts to build systems of global governance that might potentially manage humanity's relationship with the Earth system. An important milestone will hopefully turn out to have been the adoption in 2015 of the Sustainable Development Goals (SDGs) which were adopted with an explicit recognition of the fundamental and inseparable relationship of interdependence between humanity and nature. The SDG 15, in particular, aims to protect, restore and promote sustainable use of terrestrial ecosystems, sustainably manage forests, combat desertification, and halt and reverse land degradation and halt biodiversity loss. Whilst this may be the most relevant goal from the perspective of plant conservation, SDG 14 provides an essential equivalent focus on life in the oceans and many of the other goals have more than passing relevance to botanic gardens (Blackmore, in press). Sustainable Development Goal 2, for example, focuses on food security and SDG 11 on making cities and human settlements inclusive, safe, resilient and sustainable. Both target issues are addressed by many, though not all, botanic gardens.

The SDGs are an important signal of a growing international consensus focused on action to secure a sustainable biosphere. It is too early to predict whether or not they will mark a significant turning point but they are complex and subtle enough to capture the interconnectedness that was lacking in the earlier Millennium Development Goals agreed in 2000. Furthermore, the UNFCCC Paris Agreement in 2015 stated that it was 'recognising that climate change represents an urgent and potentially irreversible threat to human societies and the planet and thus requires the widest possible cooperation by all countries'. This clearly indicates the intention to formulate plans of action that transcend narrow national agendas and aim for a better future of the kind envisaged in the SDGs.

Within the scientific community there has been debate about whether the biosphere has been damaged to the extent that we must inevitably face a bleak future or whether it might be possible to achieve a 'good' Anthropocene. Socolow (2015) proposes that we should focus on destiny and identify those actions that would secure the best outcomes for the future. The numerous indicators associated with each of the SDGs (UNDP, 2015) can be regarded as measures of progress towards defining a destiny of choice for humanity and nature, rather than an unwelcome default future determined by our lack of action. The agenda they define will shape the priorities and strategic plans of many kinds of organisations, including botanic gardens (Blackmore, in press). Only time will tell whether international efforts to reverse the declining trends evident in so many indicators will succeed, but in the meantime it surely makes sense to intensify our efforts. Making a positive contribution to tackling the global issues of our time is now at the heart of the strategies of many governmental and non-governmental organisations around the world.

1.3 Botanic Gardens and Global Change

This book focuses on plant conservation and, in particular, the work of botanic gardens. Plant conservation in its widest sense, and with comprehensive application, is essential to achieve the SDGs and we are convinced that botanic gardens can and should play a leading role in transforming our world. Botanic gardens are key stakeholders in a sustainable future because of their unique potential to help maintain the diversity of the photosynthetic base of the food chain and cycles in nature. Their institutional capacity to contribute to such profound issues stems from the living and preserved collections of plant diversity they develop and curate, coupled with their expertise in plant taxonomy, identification, propagation and cultivation (for reviews of the work of botanic gardens see, for example, Havens *et al.*, 2006, Maunder, 2008, Chen *et al.*, 2009, Donaldson, 2009, Oldfield, 2010b, Aronson, 2014, Blackmore, in press).

As the biodiversity conservation agenda developed during the twentieth century, the conservation of plant species generally received less attention, sympathy and resources than animal conservation. In the absence of popular movements calling for global plant conservation, botanic gardens have coordinated much of the progress in this arena. For over five centuries botanic gardens have been major centres for the scientific study of plant diversity. Originally focusing on medicinal plants and later the development of tropical crops, botanic gardens began their global collaboration in the conservation of plant diversity in the 1970s. Two international conservation conferences on this theme were held at the Royal Botanic Gardens (RBG), Kew, in 1975 and 1978 and IUCN's Survival Service Commission set up the Threatened Plants Committee (TPC) in 1974 with its Secretariat based at Kew. A statement from the second Kew conference noted that 'Conserving the world's flora is far too great and diverse to be carried out by any one botanic garden, any one country or even in any one continent. Co-operation, co-ordination and partnerships are essential.'

The Botanic Gardens Conservation Co-ordinating Body (BGCCB) was subsequently formed in 1979 under the auspices of IUCN with BGCCB conceived as part of the Threatened Plants Secretariat based at Kew. The initial aims of the BGCCB were: to find out which plants identified as rare by IUCN are in cultivation and to keep members in touch and promote cooperation through a newsletter. In 1984, IUCN and WWF jointly launched a Plant Conservation Campaign and Programme – 'To save the plants that save us.' One of the objectives of the Programme was to work with botanic gardens, helping them to develop their conservation role. In 1985 an IUCN Conference on botanic gardens and the World Conservation Strategy was held in Las Palmas, Gran Canaria. An outcome of this was the Botanic Gardens Conservation Strategy ultimately published in 1989. Resulting from a recommendation at the Las Palmas Conference, the BGCCB was redeveloped as the IUCN Botanic Gardens Conservation Secretariat (BGCS). Subsequently, BGCS became independent from IUCN and was established as a charitable organisation, Botanic Gardens Conservation International (BGCI) supported by its global membership.

In 1998, BGCI launched a process to prepare a new international strategy for botanic gardens updating the 1989 version. The International Agenda for Botanic Gardens in Conservation was published and launched in 2000 and was an important document in the development of the Global Strategy for Plant Conservation (GSPC) a ground-breaking initiative within the framework of the CBD. The GSPC resulted from a call from Professor Peter Raven at the XVI International Botanical Congress held in St Louis in 1999 with a resolution urging the world community to recognise plant conservation as an outstanding global priority. This call was in recognition of the lack of popular and political support for the conservation of plant diversity. Responding to the resolution, BGCI organised a meeting in April, 2000 held in Las Palmas de Gran Canaria, Spain with representation from leading botanic gardens together with IUCN and other national and international organisations. The results of this meeting were published as 'The Gran Canaria Declaration'. In this document the group resolved that a Global Strategy for Plant Conservation and an associated programme for its implementation should be developed urgently, within the framework of the CBD. The Gran Canaria Declaration was submitted to the Fifth Meeting of the Conference of the Parties (COP5) to the CBD, held in Nairobi, Kenya, in May 2000, by Brazil and Colombia. Subsequently, following a series of consultations convened by the CBD, a Strategy was prepared and in 2002, the GSPC was adopted unanimously at CBD COP6 in The Hague, the Netherlands (Wyse Jackson and Kennedy 2009). The GSPC includes 16 output-oriented targets covering all aspects of the conservation and sustainable use of plants. In 2010, an updated version of the GSPC was agreed with the 16 targets modified to reflect the growing awareness of the impacts of climate change on plant diversity. Its implementation links to the overall UN Strategic Plan for Biodiversity 2011–2020 with its CBD Aichi Targets agreed in 2010. These are in turn reflected in the SDGs which were discussed earlier.

Box 1.2 *The Global Strategy for Plant Conservation Targets 2011–2020*

Objective I: Plant diversity is well understood, documented and recognised

Target 1: An online flora of all known plants.

Target 2: An assessment of the conservation status of all known plant species, as far as possible, to guide conservation action.

Target 3: Information, research and associated outputs, and methods necessary to implement the Strategy developed and shared.

Objective II: Plant diversity is urgently and effectively conserved

Target 4: At least 15 per cent of each ecological region or vegetation type secured through effective management and/or restoration.

Target 5: At least 75 per cent of the most important areas for plant diversity of each ecological region protected with effective management in place for conserving plants and their genetic diversity.

Target 6: At least 75 per cent of production lands in each sector managed sustainably, consistent with the conservation of plant diversity.

Target 7: At least 75 per cent of known threatened plant species conserved *in situ*.
Target 8: At least 75 per cent of threatened plant species in *ex situ* collections, preferably in the country of origin, and at least 20 per cent available for recovery and restoration programmes.
Target 9: Seventy per cent of the genetic diversity of crops including their wild relatives and other socio-economically valuable plant species conserved, while respecting, preserving and maintaining associated indigenous and local knowledge.
Target 10: Effective management plans in place to prevent new biological invasions and to manage important areas for plant diversity that are invaded.

Objective III: Plant diversity is used in a sustainable and equitable manner
Target 11: No species of wild flora endangered by international trade.
Target 12: All wild harvested plant-based products sourced sustainably.
Target 13: Indigenous and local knowledge innovations and practices associated with plant resources maintained or increased, as appropriate, to support customary use, sustainable livelihoods, local food security and health care.

Objective IV: Education and awareness about plant diversity, its role in sustainable livelihoods and importance to all life on Earth is promoted
Target 14: The importance of plant diversity and the need for its conservation incorporated into communication, education and public awareness programmes.

Objective V: The capacities and public engagement necessary to implement the strategy have been developed
Target 15: The number of trained people working with appropriate facilities sufficient according to national needs, to achieve the targets of this Strategy.
Target 16: Institutions, networks and partnerships for plant conservation established or strengthened at national, regional and international levels to achieve the targets of this Strategy.

Global progress in implementation of the GSPC was reviewed in 2014 (Sharrock *et al.*, 2014) and work towards the various targets is a theme running throughout the book. As the *Plant Conservation Report 2014* reminds us,

Plants are essential for all life on earth. The uptake of carbon dioxide, one of the principle greenhouse gases, during photosynthesis is the major pathway by which carbon is removed from the atmosphere and made available to humans and animals for growth and development. Plant diversity also underpins all terrestrial ecosystems and these provide the basic life-support systems on which all life depends. *(Sharrock* et al.*, 2014)*

And yet, as the same report notes, 'Today, we continue to lose plant diversity with almost casual disregard. We cannot afford to watch from the side-lines.'

To create the more equal and sustainable world envisaged in the post-2015 development agenda and the 2050 vision of the Strategic Plan for Biodiversity 2011–2020 requires us 'to invest in securing continuity in the silent work of plants'. Further calls for action have followed. In 2014, a conference of the UNESCO Man and the Biosphere (MAB) programme called for the mainstreaming of the sustainable use and conservation of plants into national and local sustainable development strategies and for the necessary resources to support the full spectrum of botanical institutions (Rakotoarisoa *et al.*, 2015). A recent review by Corlett (2016a) also underscores the importance of plant diversity in a changing world and makes the point that, given sufficient resources and effort, there is no need for any more plant extinctions.

Against this background, we see the scope of this book as encompassing some of the most pressing challenges of our times and concerned with arguably the most important interventions that can be made for the future of both the biosphere and humanity. This may come as a surprise to some, because botanic gardens are often misunderstood as simply superior parks, with gardening perceived as an activity that can be undertaken by anyone. The origins of such misconceptions are easily understood. Botanic gardens are amongst the most aesthetically pleasing of all designed landscapes and tending plants in a domestic setting is an almost universal experience. We hope that this book will help to make the case that the increasingly strategic global community of over 3000 botanic gardens can constitute a major force for conservation action leading to positive change and a sustainable future (see also, Donaldson, 2009; Oldfield, 2009; Blackmore, in press).

Botanic gardens and the wider plant conservation community already have much of the know-how to secure maximum plant biodiversity for the future, although the distribution of capacity is uneven around the world and there is a need for significantly greater coordination and sharing of efforts. As chapters in this book demonstrate, botanic gardens have expertise in plant taxonomy and systematics, the key to knowing how to find and identify plants in the wild, how to bring them into cultivation in living collections (or to secure them in seed banks), how to propagate them, and to re-establish them back into the wild. Botanic gardens also provide excellent locations for showcasing the diversity of plants, enabling education, involvement and public outreach. Thus botanic gardens, together with arboreta, seed banks and other closely related institutions, occupy a distinctive niche. Of course, parts of their distinctive mix of skills and activities overlap with those of other sectors and organisations. Agriculture and forestry are predicated on an essentially similar set of skills, directed towards specific plants: those known to be immediately and directly useful to humanity for their products, from timber, to fibres, food, medicines and biofuels. These are, of course, the plants we depend upon most in our daily lives and so the attention they receive from forestry and agriculture organisations as well as botanic gardens is more than justified. Botanic gardens are unique in having collections of the many species not known to be of immediate utility to humankind. They have, therefore, the potential to embrace the entire plant kingdom. We are convinced that there are no useless plants and consider inadequate a perspective that regards the richness of nature as merely a resource for our benefit and so fails to see the wider connections in the web of life. Some have suggested that there is redundancy in nature and that we only need to preserve a proportion of species diversity to meet our future needs. In this, we agree with Corlett (2016a) and others who have argued that, even if there is functional redundancy, preserving maximum diversity is the best insurance policy in a time of rapid global change. The present trajectory of our planet and its habitants will increasingly test the resilience and adaptability of the biosphere. It is therefore prudent to follow the precautionary principle and aim to keep the maximum possible biodiversity at all three levels recognised and addressed by the CBD: the ecosystem level, species level and genetic level. As Aldo Leopold wrote in 1949, 'To keep every cog and wheel is the first precaution of intelligent tinkering' (Leopold, 1968). This idea, together with the intention to preserve species in protected areas where human impact has been minimised, became fundamental principles of conservation biology as defined by Soulé (1985) and others.

In recent years a vigorous debate has sprung up about the purposes and arenas of conservation in a rapidly changing world (see Kloor, 2015 for an excellent summary). The perspective put forward by Kareiva and Marvier (2012) is that conservation in the Anthropocene may not be able to rely on protected areas and traditional approaches in the face of climate change on a planet dominated by humanity which is already seeing the development of novel ecosystems. They argue that, in conservation, people must be taken into account, as well as nature. They place strong emphasis on ecosystem services, as indeed do the principles of implementation for the CBD. The two sides of this debate are not mutually exclusive and reconciliation between them is not only possible but necessary, as Tallis and Lubchenco (2014) point out. The role of humanity in creating and maintaining altered ecosystems has rightly attracted much attention (see, for example, Hobbs *et al.*, 2013, Corlett 2016b) since the concept of novel ecosystems (Chapin and Starfield, 1997) was introduced. Morse *et al.* (2014) explored and reviewed the discussions that have surrounded novel ecosystems, proposing revised definitions and a framework based on natural ecosystems, impacted ecosystems, designed ecosystems and novel ecosystems. They argued for a narrower definition of the novel ecosystem concept, excluding large-scale anthropogenic impacts such as climate change. However, just as controversy continues over the fundamental goals of conservation, so it does over the novel ecosystem concept. Murcia *et al.* (2014) provide an important critique of the concept, arguing for the importance of the emerging field of ecological restoration (see, for example, Aronson, 2014). It can be anticipated that further discussion will follow, not only on conceptual issues and policy but also on how and when to intervene. Kueffer and Kaiser-Bunbury (2014) make a compelling case that 'in human-dominated landscapes, conservation depends on reconciling conflicting concepts; preserving the qualities of historical (or pristine) nature will rely on human design, and novel ecosystems will dominate wildlands'. This approach provides a logical way forward that can be applied even in highly modified vegetation, such as that found in the granitic islands of the Seychelles (Kueffer *et al.*, 2013).

Protected areas are likely to continue to preserve the best examples of species rich biodiversity and some of the world's most beautiful and important landscapes. Heywood (2015) asked whether *in situ* conservation of threatened plants might be an unattainable goal, and provides a detailed review of the effectiveness of protected areas for plant conservation, noting many practical shortcomings. Nevertheless, protected areas should continue to be central to global efforts in conservation. However, given the shuffling effect of climate change on species assemblages even protected areas may increasingly need proactive management, including the assisted migration of species. Furthermore, as Kareiva and Marvier (2012) have argued we need to find ways in which biodiversity can be preserved in urban environments, agricultural landscapes and other places inherently less biodiversity rich than protected areas generally are. One area of growing importance is urban biodiversity. With the rise of new megacities, especially in Asia, attention is increasingly being placed on creating sustainable cities in accordance with SDG 11 (see, for example, Elmqvist *et al.*, 2013). The concept of 'designer ecosystems' (see, for example, Ross *et al.*, 2015) is of growing importance, especially in the least protected, or most damaged, landscapes. Clearly, the broadest possible definition of the scope of conservation will make the greatest contribution to a sustainable future.

With this in mind, we view the overall goal of conservation as obtaining the best possible outcome for biodiversity everywhere, at ecosystem, species and genetic levels. One important consideration arising from this perspective is that it is not just rare plants that need our attention. Of course it makes sense to prioritise efforts to save the very rarest plants including those reduced to a small and vulnerable population size, as is being done in China through the Conservation Programme for Wild Plants with Extremely Small Populations in China (Ren *et al.*, 2012). Pragmatically we may have to make choices and as noted by Oldfield, 2010a, arguably, the people who live closest to threatened plant species and whose livelihoods are dependent on plant diversity should make the decisions. Another important consideration is the special importance of tree species (Oldfield, 2009: Newton *et al.*, 2015; Cavender *et al.*, 2015) which provide the structural framework for so much of the planet's terrestrial biodiversity. Furthermore, too many once common plant species have seen significant declines around the world without necessarily facing the threat of either local or global extinction. This progressive and often unobserved erosion of biodiversity is important, especially because we generally know very little about the loss of genetic diversity it entails. It is well known that resilience to climate change is enhanced by the genetic diversity present in widespread species (see, for example, Corlett 2017; Hollingsworth *et al.*, Chapter 2 this volume). Recent studies have shown, either by modelling (see, for example, Cunze *et al.*, 2013) or by field observation (Engler *et al.*, 2011), that the natural migration rates of plant species are unlikely to keep pace with climate change. Such impacts have also been modelled at the continental scale (see, for example, McLean *et al.*, 2005) and suggest that profound changes are likely in regional floristic diversity.

It should also be noted that the wild populations of some species face severe and targeted impacts often despite being surrounded by relatively stable and secure vegetation. What such species usually have in common is the existence of market demand that drives over-harvesting in the wild. The variety of plant species threatened in this way is surprisingly broad, encompassing orchids, succulents, cycads and other plants favoured by horticultural collectors some of whom desire wild-collected plants regardless of the impact this has on already threatened species. Medicinal plants suffer the same problems including the existence of a preference for wild-harvested rather than cultivated material. Much of this overexploitation is illegal, with wildlife and environmental crime now operating on a vast scale (Nellemann *et al.*, 2014). Timber is the third largest commodity in international trade with illegal logging the most economically significant environmental crime which impacts on a wide range of species. Perhaps the most striking examples (see Chapter 3) are the extremely high prices paid for rosewood (*Dalbergia* and *Pterocarpus* species) with much of the extraction and trade in contravention of national and international law (see, for example, Ratsimbazafy *et al.*, 2016).

From this introduction it will be apparent that the threats to plant diversity are numerous and varied and that solving these challenges is a fundamental prerequisite of maintaining the full diversity of plants which have shaped and maintained the environments in which we live. It is also important to note that the taxonomic inventory of plant diversity is currently far from complete and that botanic gardens are important institutions in finding the as yet undescribed 'missing species' (see, for example, Bebber *et al.*, 2010). There have been several recent estimates of the total number of flowering plant species (see Pimm and Joppa, 2015, for a review), with Pimm and Joppa (2015) concluding that there are probably about 450,000 flowering plant species (Joppa *et al.*, 2011), 10–20 per cent of which await discovery and description, and 27–30 per cent of flowering plants are threatened with extinction. Addressing the question, 'Can we name Earth's species before they go extinct,' Costello *et al.* (2013) estimate that there are 5 ± 3 million species on Earth, of which 1.5 million have so far been named. Given that over 18,000 new species (in all taxonomic groups) are being described each year they argue that most will have been described by 2020 if the total number of species is 5 million. These and other issues are explored more deeply in the chapters which follow.

Chapter 2 considers the contribution of botanic gardens and related institutions to the exploration and documentation of biodiversity as a fundamental prerequisite for the conservation of plants. It explores the application of the latest genetic techniques to research focused at the level of the plant species and looks at the relevance of new approaches to the field of plant conservation. Our current knowledge of plant diversity including its biogeographical distribution is the focus of Chapter 2 providing an overview of the global resource that we need to conserve. This chapter discusses how advances in bioinformatics are enabling data sharing and analysis on an unprecedented scale. The rapidly expanding field of conservation genetics and the importance of integrating genetic information into conservation programmes are emphasised. We know that it is important to conserve as much genetic diversity as possible for each species, but how in practice can we do this?

The nature of the major threats to plant species and plant diversity are described in Chapter 3, together with an account of the international systems employed to record the status of plants in the wild and some of the important initiatives that respond to these threats. The IUCN Red List system and other systems such as the US G ranking and federal listing system are described. The Sampled Red List Index for plants undertaken by RBG, Kew and the Natural History Museum, London, indicated that 22 per cent of the world's plants are threatened with extinction (Royal Botanic Gardens, 2012). For some groups the proportion of threatened species is much higher, for example, 63 per cent of cycad species are threatened in the wild.

Chapter 4 focuses on *in situ* conservation by considering progress in plant exploration, plant inventory techniques, the identification of priority areas for *in situ* conservation, assessing important plant areas using GIS, ethnobotanical studies and the development of different forms of *in situ* protection. The role of tropical botanic gardens at the front line of *in situ* land management in the tropics (Chen *et al.*, 2009) and the necessity for partnerships with forestry organisations and local communities are explored.

The *ex situ* conservation of plant species in botanic gardens, arboreta and seed banks is the theme of Chapter 5. Such *ex situ* collections of living plants currently hold at least 170,000 species a figure approaching half of the known total. The legislative framework within which this work is carried out is explored and recent initiatives, such as the Global Seed Conservation Challenge, are introduced. The special challenges presented by the many 'exceptional species' which cannot be preserved in seed banks are explained and current efforts to establish alternatives such as cryo-preservation described.

The increasing involvement of botanic gardens in ecological restoration is the subject of Chapter 6 which shows how the theory of ecological restoration is being put into practice utilising the unique skill sets of botanic gardens. The successes and failures of single species reintroductions and the challenges in scaling up to the landscape level are considered. The role of botanic gardens in the assisted migration of endangered species is and the central importance of botanic garden horticultural techniques.

Chapter 7 explores the role of botanic gardens in other innovative initiatives providing potential solutions for global problems including climate change mitigation and adaption, fuel shortages and food security. Topics addressed include the role of plant diversity-driven research into C_4 photosynthesis, root nodule symbiosis, the importance of crop wild relatives and research into biofuels. It is clear that despite centuries of exploring the diversity of plants and their properties, we have much to discover and that the latest techniques of plant science have the potential to resolve issues of profound importance to a sustainable future.

That such a future requires the active engagement of citizens in wider society could not be in doubt. However, as Chapter 8 considers, there is a growing disconnection between people and nature. The unique role of botanic gardens in enabling safe access to nature, an appreciation of plants and reconnecting an increasingly urbanised world with plant diversity is covered. Examples are given of the ways in which botanic gardens are serving hard to reach audiences and involving people from 'cradle to grave' through innovative approaches. The programmes described are intended to influence and change behaviour, answering the question, what part can botanic gardens play in changing society?

Chapter 9 considers the outcomes and impact of the plant conservation efforts of botanic gardens and the wider plant conservation community particularly in the concept of the GSPC.

The final chapter, Chapter 10, provides a brief summary of the key emerging themes from the previous chapters and focuses on the main knowledge gaps, and how future research might address them. The future priorities for the botanic garden community are explored together with the need for integration of botanic garden work with the broader conservation agendas of many other agencies.

Despite the gravity of the biodiversity crisis and the overarching threat of climate change we hope the book presents a positive perspective, focusing on the numerous ways in which it is possible to tackle these challenges. Botanic gardens have already proven their current value and future potential to contribute to plant conservation, a key foundation for a sustainable future. Now there is an urgent necessity for botanic gardens to accelerate and scale up their activities in order achieve the necessary impact. If they can win support from political leaders and wider society, botanic gardens can, as the contents of this book affirm, be amongst the most significant contributors to the Future We Want.

Note

1. See http://www.fao.org/docrep/015/i2624e/i2624e00.htm

References

Adams, W. M. (2004). *Against Extinction: The Story of Conservation*. London: Earthscan.
Aronson, J. (2014). The Ecological Restoration Alliance of Botanic Gardens: a new initiative takes root. *Restoration Ecology*, 22: 713–715.
Bebber, D. P., Carine, M. A., Wood, J. R. I. et al. (2010). Herbaria are a major frontier for species discovery. *Proceedings of the National Academy of Sciences*, 107 (51): 22169–22171, doi: 10.1073/pnas.1011841108.
Blackmore, S. (2009). *Gardening the Earth: Gateways to a Sustainable Future*. Edinburgh: Royal Botanic Garden.
Blackmore, S. (2017). The Future Role of Botanic Gardens. In: Friis, I. and Balslev, H. (Eds), *Collections of Tropical Plants: Legacies from the Past or Essential Tools for the Future?*
Brook, B. W., Ellis, E. C., Perring, M. P., Mackay, A. W. and Blomqvist, L. (2012). Does the terrestrial biosphere have planetary tipping points? *TREE*, 1664: 1–6.
Carson, R. (1962). *Silent Spring*. Boston, MA: Houghton Mifflin.
Cavender, N., Westwood, M., Bechtoldt, C. et al. (2015). Strengthening the conservation value of *ex situ* tree collections. *Oryx*, 49(3): 416–424.
Chapin, F. S. III and Starfield, A. M. (1997). Time lags and novel ecosystems in response to transient climatic change in Arctic Alaska. *Climatic Change*, 3: 449–461.

Chen, J., Cannon, C. H. and Hu, H. B. (2009). Tropical botanical gardens: at the *in situ* ecosystem management frontier. *Trends in Plant Science*, 14: 584–589.

Corlett, R. T. (2016a). Plant diversity in a changing world: status, trends and conservation needs. *Plant Diversity*, 1: 11–18.

Corlett, R. T. (2016b). Restoration, reintroduction and rewilding in a changing world. *Trends in Ecology and Evolution*, 31(6): 453–462.

Corlett, R. T. (2017). A bigger toolbox: biotechnology in biodiversity conservation. *Trends in Biotechnology*, 35(1): 55–65, doi: http://dx.doi.org/10.1016/j.tibtech.2016.06.009

Costello, M. J., May, R. M. and Stork, N. E. (2013). Can we name Earth's species before they go extinct? *Science*, 329: 413–416.

Crutzen, P. J. (2002). Geology of mankind. *Nature*, 415: 23.

Cunze, S., Heydel, F. and Tackenberg, O. (2013). Are plant species able to keep pace with a rapidly changing climate? *PLoS ONE*, 8(7): doi: http://dx.doi.org/10.1371/journal.pone.0067909.

Díaz, S. and Cabido, M. (2001). Vive la différence: plant functional diversity matters to ecosystem processes. *Trends in Ecology and Evolution*, 16: 646–655.

Díaz, S., Fargione, J., Chapin III, F. S. and Tilman, D. (2006). Biodiversity loss threatens human well-being. *PLoS Biology*. 4: 1300–1305.

Donaldson, J. S. (2009). Botanic gardens science for conservation and global change. *Trends in Plant Science*, 14: 608–613.

Ellis, E. C. (2011). Anthropogenic transformation of the terrestrial biosphere. *Philosophical Transactions of the Royal Society*, 369: 1010–1035.

Ellis, E. C., Kaplan, J. O., Fuller, D. Q., Vavrus, S., Goldewijk, K. K. and Verburg, P. H. (2013). Used planet: a global history. *PNAS*, 110: 7978–7985.

Elmqvist, T., Fragkias, M., Goodness, J. *et al.* (2013). *Urbanisation, Biodiversity and Ecosystem Services: Challenges and Opportunities*. The Netherlands: Springer, 743 pp., doi: 10.1007/978-94-007-7088-1.

Engler, R., Randin, C. F., Thuiller, W. *et al.* (2011). 21st Century climate change threatens mountain flora unequally across Europe. *Global Change Biology*, 17: 2330–2341.

Havens, K., Vitt, P., Maunder, M., Guerrant, E. O. and Dixon, K. (2006). *Ex-situ* plant conservation and beyond. *BioScience*, 56: 525–531.

Heywood, V. H. (2015). *In situ* conservation of plant species: an unattainable goal? *Israel Journal of Plant Sciences*, 1–21, DOI: 10.1080/07929978.2015.1035605.

Hobbs, R. J., Higgs, E. S. and Hall, C. (2013). *Novel Ecosystems: Intervening in the New Ecological World Order*. Chichester: John Wiley & Sons, 380 pp.

Joppa, L. N., Roberts, D. L. and Pimm, S. L. (2011). How many species of flowering plants are there? *Proceedings of the Royal Society B*, 278: 554–559.

Kareiva, P. and Marvier, M. (2012). What is conservation science? *BioScience*, 62: 962–969.

Kloor, K. (2015). The battle for the soul of conservation science. *Issues in Science and Technology*, 31: 74–79.

Kueffer, C. and Kaiser-Bunbury, C. N. (2014). Reconciling conflicting perspectives for biodiversity conservation in the Anthropocene. *Frontiers in Ecology and the Environment*, 12: 131–137.

Kueffer, C., Beaver, K. and Mougal, J. (2013). Case Study: Management of Novel Ecosystems in the Seychelles. In: Hobbs, R. J., Higgs, E. S. and Hall, C. (Eds), *Novel Ecosystems: Intervening in the New Ecological World Order*. Chichester: John Wiley & Sons, pp. 228–238.

Leopold, A. (1968). *A Sand County Almanac and Sketches Here and There* (enlarged edition). New York: Oxford University Press.

Lovelock, J. (1988). *The Ages of Gaia: A Biography of Our Living Earth*. Oxford: Oxford University Press.

Lovelock, J. (2006). *The Revenge of Gaia: Why the Earth is Fighting Back – and How We Can Still Save Humanity*. London: Allen Lane.

Mace, G. M., Reyers, B., Alkemade, R. et al. (2014). Approaches to defining a planet boundary for biodiversity. *Global Environmental Change*, 28: 289–297.

Marsh, G. P. (1864). *Man and Nature: Or Physical Geography as Modified by Human Action*. New York: Scribner.

Maunder, M. (2008). Beyond the greenhouse. *Nature*, 455: 596–597.

McLean, C. J., Lovett, J. C., Küper, W. et al. (2005). African plant diversity and climate change. *Annals of the Missouri Botanical Garden*, 92: 139–152.

Mooney, H. A. (2010). The ecosystem-service chain and the biological diversity crisis. *Philosophical Transactions of the Royal Society B*, 365: 31–39.

Morse, N. B., Pellissier, P. A., Cianciola, E. N. et al. (2014). Novel ecosystems in the Anthropocene: a revision of the novel ecosystems concept for pragmatic applications. *Ecology and Society*, 19: 12, doi: http://dx.doi.org/10.5751/ES-06192-190212.

Muir, J. (1911). *My First Summer in the Sierra*. Boston, MA: Houghton Mifflin.

Muir, J. (1992). *John Muir: The Eight Wilderness Discovery Books*. London: Hodder and Stoughton.

Murcia, C., Aronson, J., Kattan, G. H., Moreno-Mateos, D., Dixon, K. and Simberloff, D. (2014). A critique of the 'novel ecosystem' concept. *Trends in Ecology and Evolution*, 29: 548–553.

Nellemann, C., Henriksen, R., Raxter, P., Ash, N. and Mrema, E. (Eds) (2014). The Environmental Crime Crisis: Threats to Sustainable Development from Illegal Exploitation and Trade in Wildlife and Forest Resources. *A UNEP Rapid Response Assessment*. Nairobi and Arendal: United Nations Environment Programme and GRID-Arendal.

Newton, A., Oldfield, S., Rivers, M. et al. (2015). Towards a global tree assessment. *Oryx*, 49(3): 410–415.

Oldfield, S. F. (2009). Botanic gardens and the conservation of tree species. *Trends in Plant Science*, 14: 581–583.

Oldfield, S. F. (2010a). Plant conservation: facing tough choices. *BioScience*, 60(10): 778–779.

Oldfield, S. F. (2010b). *Botanic Gardens: Modern-Day Arks*. London: New Holland.

Pimm, S. L. and Joppa, L. N. (2015). How many plant species are there, where are they, and at what rate are they going extinct? *Annals of the Missouri Botanical Garden*, 100: 170–176.

Rakotoarisoa, N. R., Blackmore, S. and Riera, B. (2015). *Botanists of the 21st Century: Roles, Challenges and Opportunities*. Paris: UNESCO.

Ratsimbazafy, C., Newton, D. J. and Stéphane, R. (2016). *Timber island: Rosewood and Ebony trade of Madagascar.* Cambridge, UK: TRAFFIC.

Ren, H., Zhang, Q., Lu, H., *et al.* (2012). Wild plant species with extremely small populations require conservation and reintroduction in China. *Ambio,* 4: 913–917.

Rockström, J., Steffen, W., Noone, K. *et al.* (2009). A safe operating space for humanity. *Nature,* 461: 472–475.

Ross, M. R. V., Bernhards, E. S., Doyle, M. W. and Heffernan, J. B. (2015). Designer ecosystems: incorporating design approaches into applied ecology. *Annual Review of Enviroment and Resources,* 40: 419–413.

Royal Botanic Gardens (2012). *Plants under Pressure: A Global Assessment. IUCN Sampled Red List Index for Plants.* Richmond, UK: Royal Botanic Gardens, Kew.

Sharrock, S., Oldfield, S. and Wilson, O. (2014). *Plant Conservation Report 2014: A Review of Progress in Implementation of the Global Strategy for Plant Conservation 2011–2020.* Richmond, UK: Secretariat of the Convention on Biological Diversity, Montréal, Canada and Botanic Gardens Conservation International, Technical Series No. 81, 56 pp.

Socolow, R. (2015). Climate change and destiny studies: creating our near and far futures. *Bulletin of the Atomic Scientists,* 71: 18–28.

Soulé, M. E. (1985). What is conservation biology? *Bioscience,* 35: 727–734.

Steffen, W., Crutzen, P. J. and McNeill, J. (2007). The Anthropocene: are humans now overwhelming the great forces of nature? *Ambio,* 36: 614–621.

Steffen, W., Grinevald, J., Crutzen, P. J. and McNeill, J. (2011). The Anthropocene: conceptual and historical perspectives. *Philosophical Transactions of the Royal Society A,* 369: 842–867.

Steffen, W., Richardson, K., Rockström, J. *et al.* (2015). Planetary boundaries: guiding human development on a changing planet. *Science,* 347(6223): 736, doi: 10.1126/science.1259855.

Tallis, H. and Lubchenco, J. (2014). Working together: a call for inclusive conservation. *Nature,* 515: 27–28.

United Nations (1987). *Our Common Future: Brundtland Report.* Oxford: Oxford University Press, p. 204.

United Nations Development Programme (2012). *The Future We Want: Biodiversity and Ecosystems Driving Sustainable Development. United Nations Development Programme Biodiversity and Ecosystems Global Framework 2012–2020.* New York: United Nations.

United Nations Development Programme (2015). *Transforming our World: The 2030 Agenda for Sustainable Development.* New York: United Nations.

Wilson, E. O. (1992). *The Diversity of Life.* London: Allen Lane.

Wilson, E. O. (2002). *The Future of Life.* New York: Knopf.

Wyse Jackson, P. S. and Kennedy, K. (2009). The global strategy for plant conservation: a challenge and opportunity for the international community. *Trends in Plant Science,* 14: 578–580.

2 · *Using DNA Sequence Data to Enhance Understanding and Conservation of Plant Diversity at the Species Level*

PETER M. HOLLINGSWORTH, LINDA E. NEAVES AND ALEX D. TWYFORD

Understanding and conserving plant diversity is a pressing global challenge. In this chapter we explore the uses of genetics to enhance understanding of plant species diversity. Specifically we provide an overview of the use of genetic data for discriminating among plant species, ranging from broad-brush DNA barcoding studies to in-depth investigations of closely related species complexes. We highlight the types of data required to provide insights into differing levels of biological complexity and also explore future opportunities for further upscaling the use of genetic data at the species level in plants.

2.1 The Use of Genetics to Tell Species Apart

Characterisation of the world's species is an enormous undertaking. Across eukaryotic life, estimates of species numbers vary greatly but with widespread acceptance that only a fraction of the species on Earth have yet been described (Mora *et al.*, 2011). The situation is most acute for insects, where recent estimates suggest that the number of species may be as high as 10 million (Hebert *et al.*, 2016a,b). In plants, the outstanding task is more manageable, but still substantial. Recent extrapolations suggest something in the order of 350,000–450,000 species of flowering plants (Pimm and Joppa, 2015), with estimates of *c.*70,000 species still awaiting discovery and description (Bebber *et al.*, 2010). The mean time lag between species collection and species description is > 30 years, and about 50 per cent of the plant species awaiting discovery are considered to have already been collected and housed in herbaria (Bebber *et al.*,

2010). Major outstanding challenges include quantification of diversity in the species-rich understudied regions of the planet, understanding diversity in large complicated genera, and reconciliation of species accounts between different geographical regions.

Parallel to the challenge of plant species characterisation is plant identification. There is a general shortage of taxonomists and field biologists and even when experts are working on well-studied groups in well-studied regions, challenges remain when identification is required from sub-optimal material. Understanding regeneration patterns in species-rich systems requires identification of seedlings and juveniles – a non-trivial challenge in many long-lived taxa. Likewise, understanding food webs, pollination networks, below ground spatial architecture of roots, etc. all involve identification of plant parts that may lack the distinguishing characters used in taxonomic accounts.

Knowing how many species exist, where they occur, and being able to tell them apart is a critical element of conservation planning. Where there is uncertainty over species limits, conservation threats may be under- or over-estimated. Unrecognised cryptic species will be overlooked in legislation and in practical conservation programmes (e.g. multiple rare species may be mistaken for one common species). Likewise erroneously described species can lead to conservation resources being wasted on species that are not biologically meaningful (Hollingsworth, 2003).

Conservation challenges related to taxonomy also occur where there are difficulties in species identification. At a very basic level, identification is required to generate a species distribution and abundance, which informs whether a species is common or rare, threatened or not threatened. Understanding a species' ecology to inform its management is often dependent on seedlings, seeds or other plant parts being accurately identified in order to unravel ecological processes. Practical enforcement of illegal harvesting or other wildlife crime requires identification of samples to ascertain whether the plant material in question stems from a species for which trade is legal or not. In summary, discriminating among plant species can be crucial for many aspects of their conservation (see Box 2.1).

In light of the difficulties in telling species apart, and its importance for conservation, genetic data represent a useful contribution to the plant taxonomy and conservation tool kit, with key benefits being the sheer number of characters available, the intrinsic comparability of the four bases of DNA, the lack of environmental or developmental plasticity in genetic data and its applicability to small fragments or

> Box 2.1 *GSPC Target 1: An Online Flora of all Known Plants*
>
> A key need for plant conservation is an open access resource providing information on the world's plant species. This is the driver for Target 1 of GSPC which aims to make a major contribution to Aichi Target 19 (biodiversity knowledge improved, shared and applied). Specifically, this target is focused on the World Flora Online (WFO) project.[1] The World Flora Online aims to provide an open-access online compendium of the world's plant species. Its mode of operation is to provide a portal assembling existing knowledge, and to stimulate further exploration, species discovery and taxonomy on poorly known plant groups and geographic regions. The tools and techniques described in this chapter contribute towards this process.
>
> The WFO builds on *The Plant List*,[2] which was developed through international collaboration in response to the earlier phase of the GSPC (2002–2010).
>
> By early 2016, 34 institutions worldwide had joined the WFO project and agreed on the software to be used for the development of a public portal for the Flora. As with *The Plant List*, botanic gardens are playing a leading role in the development of this resource, which provides a critically important global dataset to inform conservation actions.

degraded plant material. The first few decades of research using genetics to discriminate among species made extensive use of isozymes, plastid and ribosomal restriction fragment length polymorphisms, amplified fragment length polymorphisms, and simple sequence repeats (SSRs) (Lowe *et al.*, 2004; Schlotterer, 2004). Many fascinating insights have been obtained into the nature of cryptic species, patterns of hybridisation, introgression and speciation (Arnold, 1997; Avise, 2004) and such taxonomic data has been often used to inform conservation decisions (Ennos *et al.*, 2005). However, a general property of these techniques is the disposable nature of the data. The data – the speed of movement of a band of some form (protein product, nucleic acid fragment) through a gel matrix – is poorly suited to between study comparisons, and the subsequent merging and reuse of the resulting datasets is fraught with problems.

Although some fragment assays (e.g. simple sequence repeats) continue to be widely used as proxy measures of underlying sequence variation, there has been a general move towards direct investigation of the nucleotide sequences themselves. This shift has been enabled by the increased throughput of Sanger sequencers, and new sequencing chemistries allowing the development of massively parallel high-throughput sequencing machines (Ekblom and Galindo, 2011; Twyford and Ennos, 2012). Notwithstanding important requirements of data standards, data quality and homology (Chervitz et al., 2011), genetic data can now be effectively up-scaled both in terms of the number of samples to be compared, and the number of characters recovered from individuals (Twyford, 2016). The current range of technologies thus provides greater opportunities than ever before for plant species discrimination. However, the complex nature of plant species themselves precludes simple universal technological fixes. Depending on the nature of the species/taxon/named entity in question, the tractability and data requirements for DNA-based species discrimination can vary greatly (Table 2.1).

In the following sections we explore the application of sequence-based technologies at the species level. For convenience of presentation, we make a distinction between broad-brush DNA barcoding approaches suited for large-scale screening of species encompassing broad phylogenetic diversity versus high resolution studies on complex groups using tailor-made assays for the species complex in question (Figure 2.1).

Figure 2.1 illustrates applications ranging from DNA barcoding of pollen (e.g. Case Study 2.4) and species discovery in bryophytes (e.g. Case Study 2.2), through to increasingly complex situations

Figure 2.1 Schematic illustrating the relationship between biological complexity and the number of genetic loci required. From left to right, Amazon rainforest Alberto E. Rovi; *Herbertus* David Genney; *Inga edulis* Paul Latham; switchgrass S. E. Wilco, *Mimulus guttatus* A. Twyford. For the colour version, please refer to the plate section. In some formats this figure will only appear in black and white.

Table 2.1 *Informal Summary of the Implications of Different Types of Species/Evolutionary Histories for DNA-Based Species Discrimination in Plants*

Situation	Implications for DNA-based studies	Examples
Discrete, distinct, reproductively isolated, genetically cohesive species	Taxon-specific markers relatively easy to detect, discrimination by genetic data relatively straightforward.	Hollingsworth *et al.* (1995); Bell *et al.* (2012)
Species showing marked population differentiation	Increased intra-specific divergences leading to many genetic markers being characteristic of the population, rather than the species.	Twyford *et al.* (2014)
Species showing hybridisation and/or introgression	Populations in allopatry can be relatively straightforward to tell apart, with blurred boundaries and fewer taxon-specific markers where hybridising species co-occur in sympatry.	Starr *et al.* (2013); Ruhsam *et al.* (2011)
Recent origins of species with slow mutation rates relative to divergence	Intra-specific polymorphisms may pre-date speciation reducing the likelihood of detecting discrete genetic groups.	Ruhsam *et al.* (2015); Nicholls *et al.* (2015)
Non-outcrossing species (autogams, apomicts)	Inbreeding can reduce opportunities for inter-population gene flow and species cohesion, leading to many genetic markers being characteristic of the population, rather than the species. Apomicts can differ by a small number of mutations which are difficult to detect hence different lineages may appear genetically indistinguishable.	Gornall (1999); Squirrell *et al.* (2002)

Table 2.1 (cont.)

Situation	Implications for DNA-based studies	Examples
Species of hybrid origin (e.g. allopolyploids)	Species may be distinguished by a combination of genetic markers rather than possessing unique easily detectable variants.	Ashton and Abbott (1992); Robertson et al. (2004); Robertson et al. (2010); Lu et al. (2013)
Ecotypic adaptation	Independent origins of ecotypes gives no expectation of a genetically cohesive taxon for most neutral markers. Genome scans may detect markers/genes that are associated with the ecotypes (e.g. the genomic region associated with adaptation).	Lowry (2012); Roda et al. (2013a,b); Twyford and Friedman (2015)
Taxonomically complex groups with combinations of hybridisation, breeding system polymorphism and ecotypic adaptation	Varying levels of genetic coherence of taxa: genetic markers more likely to be informative in unravelling evolutionary process, than finding discrete taxon-specific markers.	Hollingsworth 2003; Ennos et al. 2005
Species whose seeds are poorly dispersed compared to their pollen	Genetic markers inherited by seed only (e.g. plastid DNA in most angiosperms) may have low likelihood of showing taxon-specific markers and can show increased susceptibility to inter-specific introgression.	Hollingsworth et al. (2011); Naciri et al. (2012)
Hybridisation and selective sweeps	Selective sweeps can result in widespread sharing of genetic variants among distantly related species.	Percy et al. (2014); Twyford (2014)

requiring more loci (e.g. recent tropical tree radiations – Case Study 2.9 polyploidy and hybridisation – Case Study 2.10, and ecotypic adaptation – Case Studies 2.11 and 2.12).

2.2 Sequencing a Few Loci to Obtain Broad-Brush Patterns: DNA Barcoding

The principle of DNA barcoding is to use one or a few standardised regions of DNA to tell as many of the world's species apart as possible (Hebert *et al.*, 2003; Hebert *et al.*, 2016a). The use of standardised regions enables different research groups to combined efforts to build an accessible reference library of sequences as a tool for identification and also to facilitate the process of species discovery. This standardised approach has the great benefit that a common underlying reference database can be queried whether the study is focused on a particular taxonomic group, a geographic area, or a given application. In many animal groups, the use of a 648 bp portion of the mitochondrial *cytochrome oxidase* 1 gene is remarkably effective at species discrimination (Hebert *et al.*, 2016a). Sequence clusters show high concordance with morphological taxonomies, and novel sequence clusters are often indicative of previously unrecognised species (Ratnasingham and Hebert, 2013). In plants, standard DNA barcoding is based on a few plastid regions and the internal transcribed spacers of nuclear ribosomal DNA (Kress and Erickson, 2007; Hollingsworth *et al.*, 2009; Hollingsworth 2011; Hollingsworth *et al.*, 2011; Li *et al.*, 2011). The complex nature of plant species means that the approach sometimes provides species-level resolution, and other times provides resolution to a group of related species. Not surprisingly, the most successful applications of standard DNA barcoding approaches in plants have been those which match the resolving power of the technique to the question at hand (Hollingsworth *et al.*, 2016).

2.2.1 Barcoding Individuals: Specimen Barcoding

DNA barcoding individual specimens has been the mainstream use of DNA barcoding and provides insights into species limits and specimen identification. Applications range from topics as diverse as species discovery, comparative estimates of diversity, identification of plant parts, and various applied projects such as identification of invasive species, illegally traded species, and substitutions in the market place (e.g. mislabelled plant products) (Hollingsworth *et al.*, 2016).

Case Study 2.1 *Floristic Barcoding*

Establishing a comprehensive sequence reference library for an entire flora provides a baseline identification resource and an inherently comparable dataset on plant diversity to complement diversity metrics based on species counts. De Vere et al. (2012) completed the first national floristic barcoding programme for the native seed plants of Wales. The project mainly used herbarium specimens as a tissue source for 4272 DNA individuals of 1143 species, and achieved species-level discrimination in up to 75 per cent of cases, with the remainder identified to the genus level. When the data were subsampled at smaller levels (e.g. 2 km × 2 km, typical of a project investigating ecological processes at a given location) up to 93 per cent of the local species assemblage could be uniquely identified. The barcode of Wales project has been extended to cover the remainder of the UK flora, and complements many other floristic barcoding projects underway elsewhere (Hollingsworth et al., 2016). Of particular interest from a conservation perspective are floristic barcoding projects which utilise sequence data to identify areas of higher phylogenetic diversity than would be predicted from species richness estimates alone – providing new comparable metrics to support conservation planning (Shapcott et al., 2015).

Case Study 2.2 *Barcoding for Plant Species Discovery*

The identification of sequence clusters that show unexpected divergence can be indicative of the presence of cryptic taxa. Bell et al. (2012) sequenced 41 accessions of the liverwort genus *Herbertus* for four barcode markers, as part of a wider project barcoding the British liverwort flora. Samples of *Herbertus borealis* resolved as polyphyletic with specimens falling into two separate clades. Subsequent examination detected morphological and ecological differences between these two clades leading to the recognition of a new species, and recircumscription of *H. borealis*. The character-poor and relatively diminutive stature of bryophytes makes them good candidates for the presence of genetically distinct but morphologically cryptic species. Even in an extremely well-studied flora like that of the British Isles, barcoding screens of the bryophyte flora are revealing other cases of taxon discovery and taxonomic clarification (unpublished data). This is clearly relevant to the conservation of plant diversity given that cryptic species are in danger of slipping through the net and not being included in conservation planning (Hollingsworth, 2003).

Case Study 2.3 DNA Barcoding to Identify Plant Species in Trade

Biological identifications underpin product choice in the marketplace, yet many products are not readily identifiable at the point of sale. This is particularly pertinent for the huge global trade in herbal medicines and dietary supplements. *Ginkgo biloba* is frequently taken as a dietary supplement primarily targeted at brain function (memory, cognitive performance). To assess whether *Ginkgo biloba* products contained *Ginkgo biloba*, Little (2014) screened 40 *Ginkgo* preparations with a 166 bp *matK* mini barcode – with *Ginkgo* DNA recovered from 31 out of 37 products which had assayable plant DNA. Rice DNA was the most commonly encountered filler DNA when *Ginkgo* was not detected. An enhanced ability to screen the contents of dietary supplements and herbal medicines is a major step forward in monitoring an industry which has previously been difficult to regulate (Baker *et al.*, 2012; de Boer *et al.*, 2015). This general principle of identifying material that has been processed in some way is also applicable to regulation of trade in endangered species, and various studies have focused on assay development to support forensic investigations into wildlife crime, now one of the major drivers of plant extinction (e.g. *Dalbergia* spp. (Hartvig *et al.*, 2015); *Euphorbia* spp. (Aubriot *et al.*, 2013); and *Gonostylus* spp. (Ogden *et al.*, 2008)). There is great potential for increased use of DNA barcode data as evidence to inform enforcement of wildlife crime as access to forensic-standard sequencing technologies and DNA reference libraries for endangered species continue to grow.

2.2.2 Barcoding Mixtures: Metabarcoding

Massively parallel sequencing technologies enable effective identification of the species composition of mixed samples, where plant parts from different species are mixed together (Taberlet *et al.*, 2012; Creer *et al.*, 2016). Barcoding of mixtures essentially allows the identification of the previously unidentifiable, and the approach has been applied to applications as diverse as diet analysis of herbivores, assessment of the plant species diversity in pollen samples, through to surveys of insect biodiversity by sequencing the bug-splatter from a car windshield as an alternative method of sampling (Kosakovsky Pond *et al.*, 2009). The sequencing of mixtures has great potential for conservation and understanding of ecological processes, as it addresses a significant barrier to plant identification where composite mixtures of plants prevent use of conventional methods (Staats *et al.*, 2016; Hajibabaei *et al.*, 2016).

Case Study 2.4 *Barcoding Pollen*

Allergenic responses to airborne pollen are a widespread problem, with symptoms ranging from minor discomfort to more serious respiratory problems. Morphological identification of pollen grains provides some insights as to the causal species, but the lack of readily distinguishable characters across major plant families (e.g. grasses) represents a limitation on further understanding the link between individual plant species and risks to public health. DNA barcoding of airborne pollen samples offers an opportunity to at least partially address this issue. Kraaijeveld *et al.* (2015) sequenced samples from standard Burkhard pollen traps and were able to obtain genus-level resolution from barcodes, improving the typically family-level identification obtained from pollen morphology. This application of DNA barcoding to a basic societal application is an effective way of linking biodiversity to day-to-day life. In terms of conservation, enhanced identification of pollen grains (either airborne, on insects, or in sediments) represents a major step forward in understanding the ecological complexity of pollination (Bell *et al.*, 2016a,b). The reproductive biology of plants is difficult to observe, but is an essential part of their life cycle. Techniques that improve our ability to distinguish pollinators (Wong *et al.*, 2015) and pollen donors (Pornon *et al.*, 2016) present an exciting opportunity for a new optic into the cryptic process of plant reproduction.

Case Study 2.5 *Unravelling the Plant Components of Herbivore Diet*

Understanding which plant species herbivores are eating can be a non-trivial challenge if the herbivores are reclusive, wide ranging, difficult to observe or consume complex plant species mixtures. Kartinzel *et al.* (2015) used the short P6 loop of the plastid *trnL* intron to screen faecal samples from seven sympatric large mammalian herbivores from semi-arid savanna habitat in Kenya (Grevy's zebra, plains zebra, cattle, buffalo, elephant, impala, dik-dik). The study detected 110 unique plant sequence types from the faecal samples (32–62 per herbivore species) from 25 different plant families, with 70 per cent identifiable to species level and the remainder to species-group when queried against a reference library representing the local flora. The study revealed an unexpected niche differentiation in dietary preference between species, with even closely related species such as the two zebra species showing clear difference in diet. These quantifications of dietary preferences based on DNA identification of plants have clear applications for wildlife management and conservation as well as for understanding the evolutionary ecology of plant–animal interactions (Valentini *et al.*, 2009).

2.2.3 Barcoding Traces of DNA in the Environment: eDNA Barcoding

An exciting extension to metabarcoding is the ability to detect the presence of species in a given system based on traces of DNA in the environment (Bohmann *et al.*, 2014; Creer *et al.*, 2016). Such approaches offer great potential for environmental biomonitoring and are particularly amenable to standardised screening of water and soil samples (Hajibabaei, 2015; Hajibabaei *et al.*, 2016). In plants, the first applications of eDNA barcoding have focused on environmental monitoring of species/assemblages that are relatively straightforward to detect (e.g. studies summarised in Thomsen and Willerslev, 2015). eDNA approaches have been deployed effectively for monitoring rare aquatic animals (Thomsen and Willerslev, 2015) and there is clear future potential for eDNA approaches to contribute towards the search and monitoring of rare and elusive plant species of conservation importance.

Case Study 2.6 *Detecting Invasive Aquatic Plant Species from their eDNA (Environmental DNA)*

Early detection of invasive species is important for their control, yet the detection of invasive aquatic organisms in large water bodies can be challenging. As a proof of concept study, Scriver *et al.* (2015) developed taxonomically informative *matK* assays for 10 invasive alien aquatic plants representing a threat to Canadian waterbodies. The assays were then successfully tested on control eDNA samples (DNA extracted from water in which the plants were grown). Although preliminary in nature, the study illustrates the potential for biomonitoring of aquatic macrophytes via eDNA and complements similar studies on the detection of invasive fish species from eDNA traces (Mahon *et al.*, 2013).

Case Study 2.7 *Reconstructing Historical Plant Communities*

Reconstruction of past vegetation types has previously been constrained to palynological investigations of plant remains that can often be difficult to identify. Genetic analysis of eDNA in 242 ^{14}C dated sediment samples from 21 sites was used to provide new insights into past vegetational change across the Arctic over the last 50,000 years (Willerslev *et al.*, 2014). DNA sequencing of the *trnL* P6 loop showed a dominant vegetation type of dry steppe–tundra, with diversity declines between 25 and 15 kyr BP, followed by a vegetation shift to moist tundra dominated by woody plants and graminoids in the 10 kyr BP. The ongoing development of short-read sequencing technologies is perfectly suited to such studies of ancient and typically degraded DNAs as it bypasses the constraints to standard amplicon-based Sanger sequencing that were a limiting factor for earlier paleo-barcoding studies (Hagelberg *et al.*, 2015; Pedersen *et al.*, 2015; Speller *et al.*, 2016). Such studies of eDNA have the potential to track changes in the composition of plant communities in response to human-induced environmental changes (e.g. pastoral activities; Pansu *et al.*, 2015).

2.3 High Resolution Studies on Complex Groups (Sequences from Multiple Nuclear Loci)

An important fraction of plant diversity belongs to complex species groups, which defy easy taxonomic classification and understanding. Identifying species in these groups is not possible with standard broad-brush DNA barcoding approaches. To successfully study species interrelationships in these complex groups requires the development of specific assays, typically using high-resolution genomic tools. This focus on generating a large number of informative nucleotide variants (single nucleotide polymorphisms, SNPs) provides insights into relationships among species of recent origin, or those with other complex evolutionary histories (Table 2.1). These genomic assays include restriction site-associated DNA sequencing (RADseq) and related approaches (reviewed in Andrews *et al.* (2016) and capture-based approaches (reviewed in Jones and Good, 2016), which offer increasingly popular protocols for detecting thousands of SNPs across thousands of individuals. The large number of nucleotide variants will often be informative of evolutionary patterns at the population and species level, while more in-depth analysis of individual loci highly differentiated between taxa can shed light on underlying evolutionary mechanisms. This detailed genetic data may in turn be used to inform conservation actions for rare or cryptic species, or identify valuable habitats with high diversity or active evolutionary processes generating novel diversity.

2.3.1 Providing Enhanced Resolution to Understand Recent Biological Radiations

Recent speciation often gives rise to geographically restricted species, many of which are of conservation concern. Recently derived species typically show low levels of sequence divergence from congeners, making assessments of species limits and phylogenetic relationships challenging. Genomic studies of these groups must thus recover sufficient loci to discriminate between species characterised by low nucleotide divergence. Such studies can give important insights into species diversity and the mode and tempo of divergence in critical taxa and globally important biomes.

Case Study 2.8 *Resolving the* Pedicularis *Radiation in China*

A lack of informative characters obtained via Sanger sequencing of a limited number of loci has long been a limitation to evolutionary inference of recent species diversity. This issue is evident in *Pedicularis* (Orobanchaceae) where single-locus studies of the recent radiation of montane south-central China yielded low levels of resolution. Eaton and Ree (2013) revisited the evolution of this clade by analysing 40,000 RAD loci (60,000 SNPs). This resolved two major clades corresponding to the published taxonomy. They found similar levels of nucleotide divergence were recognised at the level of species in one clade and subspecies in the other, highlighting the potential utility of genomic approaches for clarifying species boundaries in complex groups, which can then be used to inform their conservation.

Case Study 2.9 *Phylogenetic Insights into a Recent Neotropical Tree Radiation*

Obtaining resolved phylogenies from recent radiations has been a persistent challenge hampering efforts to understand diversification in the tropics. This is exemplified by the neotropical legume genus *Inga*, where previous studies with 6 kb of plastid data provided limited species-level resolution. Nicholls *et al.* (2015) used a targeted capture approach to examine 264 loci, which enabled them to reconstruct a highly supported phylogeny. Although this phylogeny is at odds with the published taxonomic account, it does recover well-supported clades that correspond to other morphological groupings and thus serves to inform ongoing taxonomic revisions in the genus. The phylogeny is also being used to examine the phylogenetic distribution of herbivore defence mechanisms to assess the importance of plant–herbivore coevolution as drivers of recent diversification in the tropics.

2.3.2 Unravelling the Complexity of Polyploid Complexes

Polyploidy often causes taxonomic complexity, which in turn can create uncertainty about what to conserve (Ennos *et al.*, 2005). This is a particularly important issue for recent post-glacial floras such as those of boreal regions, where polyploidy is involved in the origin of a large proportion of the endemic species of conservation concern (Hollingsworth, 2003; Hollingsworth *et al.*, 2006; Brochmann *et al.*, 2003, 2004). Identifying taxonomic units suitable for conservation is particularly difficult in polyploid complexes, as recent polyploids may be completely reproductively isolated but show no detectable differences at conventional marker loci. Genomic data in conjunction with improved bioinformatic analyses has improved our ability to reliably score sequence differences in polyploid taxa and thus clarify the nature of species boundaries and identify taxa of conservation concern. These data can also reveal the evolutionary processes underlying the substantial diversity present in polyploid complexes.

Case Study 2.10 *Invasive Taxa, Polyploidy, and the Generation of Endemic Taxa*

Invasive species are usually considered detrimental for native floras, however the case of novel endemic taxa outside their home range is less clear-cut. Vallejo-Marin and Lye (2013) investigated the origins of an unusual monkey flower species present in the UK. Initial investigations with SSRs revealed this taxa to be a novel polyploid ($2n = 6x = 92$) derived from a hybrid of the invasive *M. guttatus* ($2n = 2x = 28$) and *M. luteus* ($2n = 4x = 44$). A larger number of SNP loci enable inference of multiple origins of the British taxon, *M. peregrinus* (Vallejo-Marín *et al.*, 2015). Whether endemic *M. peregrinus* as a species, or indeed independent origins of the species, deserve conservation status is somewhat controversial given its invasive origins. However, this example illustrates the utility of genomic tools for unravelling evolutionary processes in endemic polyploid taxa.

Case Study 2.11 *Repeated Polyploidy and Complex Genome Evolution*

Multiple rounds of genome duplication are a common feature of complex plant taxa, and this may obscure underlying evolutionary processes using standard genetic approaches. The biofuel crop switchgrass (*Panicum virgatum*) is an economically important plant that has multiple ploidy levels. Lu *et al.* (2013) studied the evolution of tetraploid (4*x*) and octoploid (8*x*) switchgrass using high-density SNP markers. The high marker number allowed them to perform stringent bioinformatic filtering to remove SNPs not confidently assigned as allelic variants of a single chromosome. Surprisingly, they found the tetraploid to be derived from the octoploid, an atypical case of large-scale genome reduction in young taxa. Although there are no direct conservation implications for switchgrass, this study highlights the potential unexpected evolutionary outcomes occurring in polyploid taxa and the power of new sequencing approaches to unravel the biology of complex groups.

2.3.3 Understanding Mechanisms of Ecotypic Adaptation

Studies of ecotypic variation can improve our understanding of the genetic basis of novel adaptations and the earliest stages of speciation (Lowry, 2012). They can also address practical challenges for conservation – namely working out whether a given morphology in a given environment represents a cohesive biological entity (protectable in legislation) or a set of independent adaptations to similar conditions without genetic coancestry (i.e. not a cohesive biological entity and not protectable by species-based legislation). There are technical difficulties in achieving this – not only due to limited nucleotide divergence, but also because the key information may be restricted to few regions of the genome. Genomic approaches are well suited to addressing this and are able to pinpoint regions of divergence in ecotypes that continue to exchange genes and conversely to infer whether ecotypes are reproductively isolated and thus better recognised at species level.

Case Study 2.12 *Genomic Analysis of Ecotypic Variation*

The taxonomy of ecotypes is particularly problematic in widespread species that grow across ecological gradients. Here, morphological differences in the field may simply reflect plasticity to environmental conditions, rather than genetic differences. Ecotypic variation within the widespread North American monkey flower *Mimulus guttatus* is a response to seasonal water availability, with the annual ecotype growing in areas that experience summer drought, and the perennial ecotype adjacent to permanent waterbodies. Their distinct morphologies suggest they could be recognised as distinct species. Twyford and Friedman (2015) showed that the majority of 38,871 SNPs are not partitioned according to ecotypes, as would be expected if there is gene flow between ecotypes. However, 276 SNPs in a single genomic region are partitioned by ecotype, and this region of divergence maintains this adaptive difference. These SNP markers are within a large chromosomal inversion, and thus point to a single genomic region maintaining ecotypic diversity in *Mimulus*. These results illustrate how genomic studies can illustrate both the nature of 'species' (in this case showing the distinct forms are not distinct species) and the evolutionary genetic mechanisms by which plants show adaptation to different environmental conditions. Such tools are now being applied to taxonomically complex groups including critically rare taxa, where there is considerable uncertainty as to what constitute species, hybrids or ecotypes (Ennos *et al.*, 2005).

Case Study 2.13 *Parallel Adaptation and Ecotypic Evolution*

The evolutionary processes underlying ecotypic variation are most complicated where multiple different ecotypes coexist across heterogeneous environments. Australian ecotypes of the *Senecio lautus* are found in diverse habitats, including sand dunes, rocky headlands and alpine habitats. RAD sequencing by Roda *et al.* (2013a) revealed that *Senecio lautus* ecotypes are not cohesive at genome-wide markers, with genetic variation partitioned by geography and not by ecotype. Further analysis of individual loci revealed ecotypic adaptation involves both novel genes and parallel recruitment of shared variants under similar environments (Roda *et al.*, 2013b). This illustrates the complex genetic mechanisms by which these genetically distinct ecotypes may evolve. This also highlights the need to conserve large pools of standing genetic variation present in widespread species and ecotypes, as this diversity may act as the raw material for the evolution of new taxa.

2.4 Future Prospects for Large-Scale Use of Sequence Data at the Species Level in Plants

2.4.1 Extending the Plant Barcode (Closing the Gap between Broad-Brush Studies and the Complexity of Plant Species)

In this chapter, we have outlined a series of DNA barcoding case studies using standard barcode approaches for telling species apart – with the well-known inherent limits to resolving power in plants. Likewise, we have outlined case studies utilising bespoke high-resolution protocols focusing on the complex biology of closely related and interrelated plant species. It is important to note that the intervening ground in this continuum is being increasingly blurred. The concept of extended DNA barcodes has been explored by Coissac et al. (2016), and projects are underway deploying genomic sequencing approaches on large-scale sample sets at floristic scales (Coissac et al., 2016). These studies are part of the wider landscape of investigations tackling the general principle of gathering large amounts of homologous sequence data over broad phylogenetic diversity (Stull et al., 2013; Li et al., 2015; Coissac et al., 2016). The rapid pace of technology development makes it difficult to predict exactly which approaches will dominate in the coming years. The complexity of plant species (see Table 2.1), make it unlikely that the semi-automated species discrimination of insects (Ratnasingham and Hebert 2013; Hebert et al. 2016b) will be feasible in plants, but it is likely that the application of 'standard protocols' (as opposed to bespoke assays) will shift towards unravelling increasing biological complexity (i.e. a 'right shift' on the x axis of Figure 2.1). How far this will go, and how cheap it will become are exciting unknowns – but an extremely positive note is that technological opportunities for using sequence data for understanding plant diversity are developing faster than ever before.

2.4.2 Developing Portable Sequencers

Conservation work is often slowed by the lengthy process of species identification. Portable field-based sequencers offer the potential for real-time identifications and insights in the field. Although this is some way from being deployable on a large scale, an important technical development is the release of the first portable next-generation sequencing platform, the Oxford Nanopore MinION. This pocket-sized device can be taken anywhere, and thus opens the possibility of field-based barcoding for extremely remote conditions. The first reports of field-based DNA identification with this technology include its use on a potential new species of frog in a rainforest in Tanzania (Hayden, 2015), with a subsequent extreme application being the first DNA sequencing in space in August 2016 (NASA, 2016). The rapid generation of data in the field has massive potential for informing biodiversity science, paralleling the application of field-based diagnostics for human disease (e.g. field-based sequencing for investigating the spread of the *Ebola* virus (Quick *et al.*, 2016)). Clearly there is enormous conservation potential of a device which would allow, for example, field-based identification of seedlings of long-lived trees, the pollen and pollinators of threatened species, the diseases that infect threatened species, or to even gain genus-level identification of every species in a complex forest system. To make such a device realistically achievable for plants in the near future, assay development is still required to focus the relatively expensive sequence output onto targeted and standardised regions of the genome to bring the per-sample cost down to affordable levels.

2.5 Concluding Remarks

Botanic gardens and natural history institutions represent critically important components of global capacity to address sustainable development via effective use of planetary resources. This is partly attributable to the centuries of effort in assembling richly annotated collections of millions of accessions of living and preserved plants as taxonomic, biomonitoring, cultural and genetic resources. It also attributable to their increasingly unique expertise reservoir of taxonomy, horticulture, biodiversity science, ethnobotany and organismal biology.

Blaxter (2016) noted that the Linnean project to document the planet's species dwarfs all other megascience projects. Technological innovation is central to this endeavour and future success will be dependent on the usual principles of prioritisation, securing resources and using them efficiently. In the context of this chapter, two concluding points are worthy of note.

- The global barcoding programme has been successful, because it involved a technology accessible to labs of varying size, scale and budgets, and it utilised an approach with limited biofunctional information which reduced access and benefit sharing (ABS) concerns, enabling movement of data and samples around the world. As standard barcodes blend towards genomic science, maintaining

accessibility to less well-resourced laboratories and adherence towards ABS concerns is essential to maintain global input to this global resource.
- Sequence data needs organismal data to contextualise it (and vice versa). The current rate-of-gain in sequencing capacity runs the risk of outstripping the process of placing this data into an interpretive context (Page, 2016). Botanic gardens and other natural history organisations have a central role in managing this flow, with continued focus on integrative approaches to organismal biology and genomic science which serve to underpin conservation and conservation science.

Acknowledgements

PMH acknowledges funding from the Scottish Government's Rural and Environment Science and Analytical Services Division (RESAS). PMH and LN acknowledge support of the Leverhulme Trust Research Project Grant RPG-2015-273. Research by ADT is supported by NERC Fellowship NE/L011336/1.

Notes

1. See www.worldfloraonline.org/.
2. See www.theplantlist.org/.

References

Andrews, K. R., Good, J. M., Miller, M. R., Luikart, G. and Hohenlohe, P. A. (2016). Harnessing the power of RADseq for ecological and evolutionary genomics. *Nature Reviews Genetics*, 17: 81–92.

Arnold, M. L. (1997). *Natural Hybridization and Evolution*. Oxford: Oxford University Press.

Ashton, P. A. and Abbott, R. J. (1992). Multiple origins and genetic diversity in the newly arisen allopolyploid species, *Senecio cambrensis* Rosser (Compositae). *Heredity*, 68: 25–32.

Aubriot, X., Lowry, P. P. II, Cruaud, C., Couloux, A. and Haevermans, T. (2013). DNA barcoding in a biodiversity hot spot: potential value for the identification of Malagasy *Euphorbia* L. listed in CITES Appendices I and II. *Molecular Ecology Resources*, 13: 57–65.

Avise, J. C. (2004). *Molecular Markers, Natural History and Evolution*, 2nd edition. Boston, MA: Sinauer.

Baker, D. A., Stevenson, D. W. and Little, D. P. (2012). DNA barcode identification of Black Cohosh herbal dietary supplements. *Journal of AOAC International*, 95: 1023–1034.

Bebber, D. P., Carine, M. A., Wood, J. R. I. *et al.* (2010). Herbaria are a major frontier for species discovery. *Proceedings of the National Academy of Sciences*, 107: 22169–22171.

Bell, D., Long, D. G., Forrest, A. D., Hollingsworth, M. L., Blom, H. H. and Hollingsworth, P. M. (2012). DNA barcoding of European *Herbertus* (Marchantiopsida, Herbertaceae) and the discovery and description of a new species. *Molecular Ecology Resources*, 12: 36–47.

Bell, K. L., Burgess, K. S., Okamoto, K. C., Aranda, R. and Brosi, B. J. (2016a). Review and future prospects for DNA barcoding methods in forensic palynology. *Forensic Science International-Genetics*, 21: 110–116.

Bell, K. L., de Vere, N., Keller, A., Richardson, R, Gous, A., Burgess, K. S. and Brosi, B. J. (2016b). Pollen DNA barcoding: current applications and future prospects. *Genome*, 59(9): 629–640.

Blaxter, M. (2016). Imagining Sisyphus happy: DNA barcoding and the unnamed majority. *Philosophical Transactions of the Royal Society B: Biological Sciences* 371: 20150329.

Bohmann, K., Evans, A., Gilbert, M. T. P. *et al.* (2014). Environmental DNA for wildlife biology and biodiversity monitoring. *Trends in Ecology and Evolution*, 29: 358–367.

Brochmann, C., Gabrielsen, T. M, Nordal, I., Landvik, J. Y. and Elven, R. (2003). Glacial survival or *tabula rasa*? The history of the North Atlantic biota revisited. *Taxon*, 52: 417–450.

Brochmann, C., Brysting, A. K., Alsos, I. G., Borgen, L., Grundt, H. H., Scheen, A.-C. and Elven, R. (2004). Polyploidy in Arctic plants. *Biological Journal of the Linnean Society*, 82: 521–536.

Chervitz, S. A., Deutsch, E. W., Field, D. *et al.* (2011). Data standards for omics data: the basis of data sharing and reuse. *Methods in Molecular Biology*, 719: 31–69.

Coissac, E., Hollingsworth, P. M., Lavergne, S. and Taberlet, P. (2016). From barcodes to genomes: extending the concept of DNA barcoding. *Molecular Ecology*, 25: 1423–1428.

Creer, S., Deiner, K., Frey, S. *et al.* (2016). The ecologist's field guide to sequence-based identification of biodiversity. *Methods in Ecology and Evolution*, 7(9): 1008–1018, doi: 10.1111/2041-1210X.12574.

de Boer, H. J., Ichim, M. C. and Newmaster, S. G. (2015). DNA barcoding and pharmacovigilance of herbal medicines. *Drug Safety*, 38: 611–620.

de Vere, N., Rich, T. C. G., Ford, C. R. *et al.* (2012). DNA barcoding the native flowering plants and conifers of Wales. *PLoS ONE*, 7: e37945.

Eaton, D. A. R. and Ree, R. H. (2013). Inferring phylogeny and introgression using RADseq data: an example from flowering plants (*Pedicularis*: Orobanchaceae). *Systematic Biology*, 62: 689–706.

Ekblom, R. and Galindo, J. (2011). Applications of next generation sequencing in molecular ecology of non-model organisms. *Heredity*, 107: 1–15.

Ennos, R. A., French, G. C. and Hollingsworth, P. M. (2005). Conserving taxonomic complexity. *Trends in Ecology and Evolution*, 20: 164–168.

Gornall, R. J. (1999). Population Genetic Structure in Agamospermous Plants. In: Hollingsworth, P. M., Bateman, R. M. and Gornall, R. J. (Eds), *Molecular Systematics and Plant Evolution*. London: Taylor and Francis, pp. 118–138.

Hagelberg, E., Hofreiter, M. and Keyser, C. (2015). Ancient DNA: the first three decades. *Philosophical Transactions of the Royal Society B: Biological Sciences*, 370: 20130371.

Hajibabaei, M. (2015). Environmental DNA barcoding: from the Arctic to the tropics – and everywhere in between. *Genome*, 58: 224–224.

Hajibabaei, M., Baird, D. J., Fahner, N. A., Beiko, R. and Golding, G. B. (2016). A new way to contemplate Darwin's tangled bank: how DNA barcodes are reconnecting biodiversity science and biomonitoring. *Philosophical Transactions of the Royal Society B: Biological Sciences*, 371: 20150330.

Hartvig, I., Czako, M., Kjaer, E. D., Nielsen, L. R. and Theilade, L. R. (2015). The use of DNA barcoding in identification and conservation of Rosewood (*Dalbergia* spp.). *PLoS ONE*, 10: e138231.

Hayden, E. C. (2015). Pint-sized DNA sequencer impresses first users. *Nature*, 521: 15–16.

Hebert, P. D. N., Cywinska, A., Ball, L. R. and deWaard, J. R. (2003). Biological identifications through DNA barcodes. *Proceedings of the Royal Society of London, series B*, 270: 313–321.

Hebert, P. D. N., Hollingsworth, P. M. and Hajibabaei, M. (2016a). From writing to reading the encyclopedia of life. *Philosophical Transactions of the Royal Society B: Biological Sciences*, 371: 20150321.

Hebert, P. D. N., Ratnasingham, S., Zakharov, E. V., et al. (2016b). Counting animal species with DNA barcodes: Canadian insects. *Philosophical Transactions of the Royal Society B: Biological Sciences*, 371: 20150333.

Hollingsworth, P. M. (2003). Taxonomic complexity, population genetics, and plant conservation in Scotland. *Botanical Journal of Scotland*, 55: 55–63.

Hollingsworth, P. M. (2011). Refining the DNA barcode for land plants. *Proceedings of the National Academy of Sciences* 108: 19451–19452.

Hollingsworth, P. M., Preston, C. D. and Gornall, R. J. (1995). Isozyme evidence for hybridization between *Potamogeton natans* and *P. nodosus* (Potamogetonaceae) in Britain. *Botanical Journal of the Linnean Society*, 117: 59–69.

Hollingsworth, P. M., Squirrell, J., Hollingsworth, M. L., Richards, A. J. and Bateman, R. M. (2006). Taxonomic Complexity, Conservation and Recurrent Origins of Self-Pollination in *Epipactis* (Orchidaceae). In: Bailey, J. P. and Ellis, R. G. (Eds), *Current Taxonomic Research on the British and European Flora*. London: BSBI, pp. 27–44.

Hollingsworth, P. M., Forrest, L. L., Spouge, J. L., Hajibabaei, M., Ratnasingham, S. and van der Bank, M. (2009). A DNA barcode for land plants. *Proceedings of the National Academy of Sciences*, 106: 12794–12797.

Hollingsworth, P. M., Graham, S. and Little, D. P. (2011). Choosing and using a plant DNA barcode. *PLoS ONE*, 6: e19254.

Hollingsworth, P. M., Li, D., VanderBank, M. and Twyford, A. D. (2016). Telling plant species apart with DNA: from barcodes to genomes. *Philosophical Transactions of the Royal Society B: Biological Sciences*, 371: 20150338.

Jones, M. R. and Good, J. M. (2016). Targeted capture in evolutionary and ecological genomics. *Molecular Ecology*, 25: 185–202.

Kartzinel, T. R., Chen, A. D., Coverdale, T. C., Erickson, D. L., Kress, W. J., Kuzmina, M. L., Rubenstein, D. I., Wang, W. and Pringle, R. M. (2015). DNA metabarcoding illuminates dietary niche partitioning by African large herbivores. *Proceedings of the National Academy of Sciences*, 112: 8019–8024.

Kosakovsky Pond, S., Wadhawan, S., Chiaromonte, F., Ananda, G., Chung, W.-Y., Taylor, J., Nekrutenko, A. and Team, T. G. (2009). Windshield splatter analysis with the Galaxy metagenomic pipeline. *Genome Research*, 19: 2144–2153.

Kraaijeveld, K., Weger, L. A., Ventayol García, M., Buermans, H., Frank, J. and Hiemstra, P. S. (2015). Efficient and sensitive identification and quantification of airborne pollen using next-generation DNA sequencing. *Molecular Ecology Resources*, 15: 8–16.

Kress, W. J. and Erickson, D. L. (2007). A two-locus global DNA barcode for land plants: the coding rbcL gene complements the non-coding trnH-psbA spacer region. *PLoS ONE*, 2(6): e508. doi:10.1371/journal.pone.0000508.

Li, D.-Z., Gao, L.-M., Li, H.-T. *et al.* (2011). Comparative analysis of a large dataset indicates that internal transcribed spacer (ITS) should be incorporated into the core barcode for seed plants. *Proceedings of the National Academy of Sciences*, 108: 19641–19646.

Li, X., Yang, Y., Henry, R. J., Rossetto, M., Wang, Y. and Chen. S. (2015). Plant DNA barcoding: from gene to genome. *Biological Reviews*, 90: 157–166.

Little, D. P. (2014). Authentication of *Ginkgo biloba* herbal dietary supplements using DNA barcoding. *Genome* 57: 513–516.

Lowe, A. J., Harris, S. A. and Ashton, P. A. (2004). *Ecological Genetics: Design Analysis and Application*. Oxford: Blackwell.

Lowry, D. B. (2012). Ecotypes and the controversy over stages in the formation of new species. *Biological Journal of the Linnean Society*, 106: 241–257.

Lu, F., Lipka, A. E., Glaubitz, J. *et al.* (2013). Switchgrass genomic diversity, ploidy, and evolution: novel insights from a network-based SNP discovery protocol. *PLoS Genetics*, 9: e1003215.

Mahon, A. R., Jerde, C. L., Galaska, M. *et al.* (2013). Validation of eDNA surveillance sensitivity for detection of asian carps in controlled and field experiments. *PLoS ONE*, 8: e58316.

Mora, C., Tittensor, D. P., Adl, S., Simpson, A. G. B. and Worm, B. (2011). How many species are there on earth and in the ocean? *PLoS Biology*, 9: e1001127.

Naciri, Y., Caetano, S. and Salamin, N. (2012). Plant DNA barcodes and the influence of gene flow. *Molecular Ecology Resources*, 12: 575–580.

NASA (2016). First DNA Sequencing in Space a Game Changer. Available online at https://www.nasa.gov/mission_pages/station/research/news/dna_sequencing [accessed 23 February 2017].

Nicholls, J. A., Pennington, R. T., Koenen, E. J. M. *et al.* (2015). Using targeted enrichment of nuclear genes to increase phylogenetic resolution in the neotropical rain forest genus *Inga* (Leguminosae: Mimosoideae). *Frontiers in Plant Science*, 6: 710.

Ogden, R., McGough, H. N., Cowan, R. S., Chua, L., Groves, M. and McEwing, R. (2008). SNP-based method for the genetic identification of ramin *Gonystylus* spp. timber and products: applied research meeting CITES enforcement needs. *Endangered Species Research*, 9: 255–261.

Page, R. D. M. (2016). DNA barcoding and taxonomy: dark taxa and dark texts. *Philosophical Transactions of the Royal Society B: Biological Sciences* 371: 20150334.

Pansu, J., Giguet-Covex, C., Ficetola, G. F. *et al.* (2015). Reconstructing long-term human impacts on plant communities: an ecological approach based on lake sediment DNA. *Molecular Ecology*, 24: 1485–1498.

Pedersen, M. W., Overballe-Petersen, S., Ermini, L. *et al.* (2015). Ancient and modern environmental DNA. *Philosophical Transactions of the Royal Society B: Biological Sciences* 370: 20130383.

Percy, D. M., Argus, G. W., Cronk, Q. C. *et al.* (2014). Understanding the spectacular failure of DNA barcoding in willows (*Salix*): Does this result from a trans-specific selective sweep? *Molecular Ecology*, 23: 4737–4756.

Pimm, S. L. and Joppa, L. N. (2015). How many plant species are there, where are they, and at what rate are they going extinct? *Annals of the Missouri Botanical Garden*, 100: 170–176.

Pornon, A., Escaravage, N., Burrus, M. *et al.*(2016). Using metabarcoding to reveal and quantify plant–pollinator interactions. *Scientific Reports*, 6: 27282.

Quick, J., Loman, N. J., Duraffour, S. *et al.* (2016). Real-time, portable genome sequencing for Ebola surveillance. *Nature*, 530: 228–232.

Ratnasingham, S. and Hebert, P. D. N. (2013). A DNA-based registry for all animal species: the Barcode Index Number (BIN) system. *PLoS ONE*, 8: e66213.

Robertson, A., Newton, A. C. and Ennos, R. A. (2004). Multiple hybrid origins, genetic diversity and population genetic structure of two endemic *Sorbus* taxa on the Isle of Arran, Scotland. *Molecular Ecology*, 13: 123–134.

Robertson, A., Rich, T. C. G., Allen, A. M. *et al.* (2010). Hybridization and polyploidy as drivers of continuing evolution and speciation in *Sorbus*. *Molecular Ecology*, 19: 1675–1690.

Roda, F., Ambrose, L., Walter, G. M. *et al.* (2013a). Genomic evidence for the parallel evolution of coastal forms in the *Senecio lautus* complex. *Molecular Ecology*, 22: 2941–2952.

Roda, F., Liu, H., Wilkinson, M. J. *et al.* (2013b). Convergence and divergence during the adaptation to similar environments by an Australian groundsel. *Evolution*, 67: 2515–2529.

Ruhsam, M., Hollingsworth, P. M. and Ennos, R. A. (2011). Genetic and phenotypic analysis of a hybrid swarm between *Geum urbanum* and *G. rivale*, plant taxa with contrasting mating systems. *Heredity*, 107: 246–255.

Ruhsam, M., Rai, H. S., Mathews, S., Ross, T. G. *et al.* (2015). Does complete plastid genome sequencing improve species discrimination and phylogenetic resolution in *Araucaria*? Molecular Ecology Resources, 15: 1067–1078.

Schlotterer, C. (2004). The evolution of molecular markers: just a matter of fashion? *Nature Reviews Genetics*, 5: 63–69.

Scriver, M., Marinich, A., Wilson, C. and Freeland, J. (2015). Development of species-specific environmental DNA (eDNA) markers for invasive aquatic plants. *Aquatic Botany*, 122: 27–31.

Shapcott, A., Forster, P. I., Guymer, G. P. et al. (2015). Mapping biodiversity and setting conservation priorities for SE Queensland's rainforests using DNA barcoding. *PLoS ONE*, 10(3): e0122164. doi:10.1371/journal.pone.0122164.

Speller, C., van den Hurk, Y., Charpentier, A. et al. (2016). Barcoding the largest animals on Earth: ongoing challenges and molecular solutions in the taxonomic identification of ancient cetaceans. *Philosophical Transactions of the Royal Society B: Biological Sciences*, 371: 20150332.

Squirrell, J., Hollingsworth, P. M., Bateman, R. M., Tebbitt, M. C. and Hollingsworth, M. L. (2002). Taxonomic complexity and breeding system transitions: conservation genetics of the *Epipactis leptochila* complex (Orchidaceae). *Molecular Ecology*, 11: 1957–1964.

Staats, M., Arulandhu, A. J., Gravendeel, B., Holst-Jensen, A., Scholtens, I., Peelen, T., Prins, T. W. and Kok, E. (2016). Advances in DNA metabarcoding for food and wildlife forensic species identification. *Analytical and Bioanalytical Chemistry* 408: 4615–4630.

Starr, T. N., Gadek, K. E., Yoder, J. B., Flatz, R. and Smith, C. I. (2013). Asymmetric hybridization and gene flow between Joshua trees (*Agavaceae*: *Yucca*) reflect differences in pollinator host specificity. *Molecular Ecology*, 22: 437–449.

Stull, G. W., Moore, M. J., Mandala, V. S. et al. (2013). A targeted enrichment strategy for massively parallel sequencing of angiosperm plastid genomes. *Applications in Plant Sciences*, 1(2): 1200497.

Taberlet, P., Coissac, E., Pompanon, F., Brochmann, C. and Willerslev, E. (2012). Towards next-generation biodiversity assessment using DNA metabarcoding. *Molecular Ecology*, 21: 2045–2050.

Thomsen, P. F. and Willerslev, E. (2015). Environmental DNA: an emerging tool in conservation for monitoring past and present biodiversity. *Biological Conservation*, 183: 4–18.

Twyford, A. D. (2014). Testing evolutionary hypotheses for DNA barcoding failure in willows. *Molecular Ecology*, 23: 4674–4676.

Twyford, A. D. (2016). Will benchtop sequencers resolve the sequencing trade-off in plant genetics? *Frontiers in Plant Science* 7: 433.

Twyford, A. D. and Ennos, R. A. (2012). Next-generation sequencing as a tool for plant ecology and evolution. *Plant Ecology and Diversity*, 5: 411–413.

Twyford, A. D. and Friedman, J. (2015). Adaptive divergence in the monkey flower *Mimulus guttatus* is maintained by a chromosomal inversion. *Evolution*, 69: 1476–1486.

Twyford, A. D., Kidner, C. A. and Ennos, R. A. (2014). Genetic differentiation and species cohesion in two widespread Central American *Begonia* species. *Heredity*, 112: 382–390.

Valentini, A., Miquel, C., Nawaz, M. A. and Bellemain, E. et al. (2009). New perspectives in diet analysis based on DNA barcoding and parallel pyrosequencing: the *trnL* approach. *Molecular Ecology Resources*, 9: 51–60.

Vallejo-Marin, M. and Lye, G. C. (2013). Hybridisation and genetic diversity in introduced *Mimulus* (Phrymaceae). *Heredity*, 110: 111–122.

Vallejo-Marín, M., Buggs, R. J. A, Cooley, A. M. and Puzey, J. R. (2015). Speciation by genome duplication: Repeated origins and genomic composition of the recently formed allopolyploid species *Mimulus peregrinus*. Evolution, 69: 1487–1500.

Willerslev, E., Davison, J. and Moora, M. *et al.* (2014). Fifty thousand years of Arctic vegetation and megafaunal diet. *Nature*, 506: 47–51.

Wong, M. M., Lim, C. L., and Wilson, J. J. (2015). DNA barcoding implicates 23 species and four orders as potential pollinators of Chinese knotweed (*Persicaria chinensis*) in Peninsular Malaysia. *Bulletin of Entomological Research*, 105: 515–520.

3 · *Conservation Assessments and Understanding the Impacts of Threats on Plant Diversity*

MALIN RIVERS

Planning for plant conservation action by botanic gardens and other conservation organisations depends on information on the natural distribution of plant species, the threats they face in the wild and the their extinction risk. Botanic gardens are a major source of such information based on both expert knowledge and written records from collecting trips, herbarium specimens, floras, keys and field guides. Significant effort has been made by some of the world's leading botanic gardens to assess the conservation status of plant species since the 1970s. It remains important to continue and increase this activity in order to reach global targets and highlight the scale of the threats faced by plants.

Knowledge of the diversity of plants remains incomplete in terms of the numbers of species and the relationships between them as discussed in Chapter 2. Plant diversity is unevenly distributed with higher concentrations of plant species found in the so-called biological hotspots which are situated mainly in tropical regions such as the Amazon, Borneo and Madagascar. These biological hotspots are not only rich in plant diversity, they are also regions with high level of threats where we are losing biodiversity at significant rates (Myers *et al.*, 2000). Due to the increasing demands of our globally expanding human population both in numbers and lifestyle expectations, there are increasing pressures on plants and their habitats.

The major threats to biodiversity and their habitats are: habitat loss, overexploitation, invasive species, pollution and climate change (Millennium Ecosystem Assessment, 2005). These threats vary in their impact and extent between different regions and ecosystems, but their common denominator is that they are all humanly driven, or at least exacerbated by humanity. Human pressure will increase as the population continues to grow, impact on the natural world increases and the

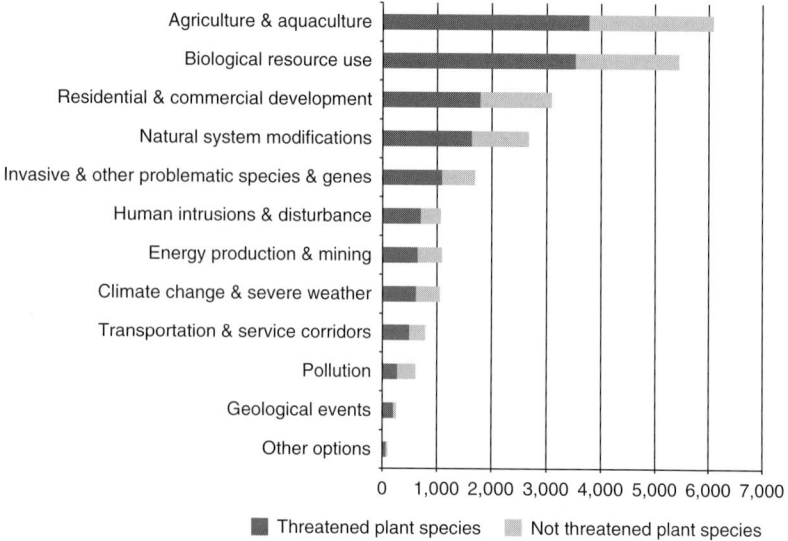

Figure 3.1 The number of plant species (threatened and not threatened) on the IUCN Red List for each major threat (Source: IUCN Red List 2016.1). For the colour version, please refer to the plate section. In some formats this figure will only appear in black and white.

land which is able to provide the commodities for our consumption decreases in extent and quality. There is no fully comprehensive analysis of which threats cause most plants to become threatened with extinction. However, using threat information submitted with the over 20,000 plant red list conservation assessments, we see that agriculture is the most common threatening process listed for plants (Figure 3.1). The second most common threat is biological resources use. This data also shows that overall the threatened plants are impacted by the same threatening processes as the non-threatened plant species. Human activities are greatly outweighing the natural threats to plant species and their habitat (Brummitt and Bachman, 2010).

The ultimate cost of these threats on plant diversity is species extinction, first on a local or regional scale and then on a global – and therefore irreversible – scale. Extinction risks due to human activities are currently estimated to be 1,000 times higher than natural background rates (De Vos et al., 2015). In order to reduce the extinction risk and ultimately biodiversity loss we need to gain better understanding about what the threats are, where they are and how to deal with them (Joppa et al., 2016).

We also need baseline information on the distribution and taxonomy of plant species so that we can monitor the impacts of various threats.

3.1 Habitat Loss

For the last 50 years, habitat and land-use change have had the biggest impact on terrestrial biodiversity. Habitat loss is the single most important threat to plant diversity, particularly in the tropics (Corlett, 2016). Habitat loss includes the total removal of habitat, as well as associated degradation and fragmentation. The majority of habitat loss is for conversion to crop land. This conversion occurs on many different scales: from small-scale slash and burn agricultural practices, to medium-scale conversion of habitat for cash crops (such as coffee and tea), to large-scale plantations of commercial crops (such as oil palm, rubber and soybean). Only regions unsuited to crop plants (deserts, boreal forest and tundra), have been less affected by land conversion (Millennium Ecosystem Assessment, 2005). Other reasons for habitat loss include conversion of land for pasture for livestock, conversion of land for mining, loss of habitat due to dam infrastructure and for coastal and urban and industrial development. In addition, even if the entire habitat is not lost or converted, habitats are degraded by impacts such as selective logging or collection of firewood – which again impacts the habitat quality and extent of many species.

Fragmentation at the landscape level is a significant factor reducing plant diversity (Kettle and Koh, 2014). Fragmentation is caused both by natural causes such as storms and fires, and also more extensively through anthropogenic land-use change such as clearing of vegetation for agriculture, or construction of roads. These practices divide and fragment larger areas of vegetation, and these smaller fragments of habitat may only be able to support smaller species populations. In addition, the edge-effect of fragmented habitats may affect species that are unable to withstand these conditions. Dispersal and pollination may also be affected as the wider ecosystem and associated species are affected. Fragmentation can also provide increased access for pests and invasive species, as well as easier access for poachers and allowing illegal collection of plants.

3.2 Overexploitation

Another major threat to plant diversity worldwide is overexploitation as mentioned in Chapter 1. Overexploitation targets certain species or groups of species. The most obvious example is the exploitation of timber trees through logging where areas are either clear-felled or specific species are targeted and extracted. To the tree species in question the felling is the main threat, but selective logging of prime timber species affects the forest and associated species by increasing fragmentation and degradation of the habitat, as well as increasing access to further exploit the environment. More specifically it creates canopy gaps and causes severe collateral damage due to construction of roads. Logging may also open up new areas to wildlife and non-timber resource exploitation, and catalyse the transition into a landscape dominated by slash-and-burn and large-scale agriculture (Reynolds and Peres, 2006).

In addition to the overexploitation of timber trees, non-timber forest products (NTFP) are also targeted. Non-timber forest products are biological resources other than timbers that are taken from forests, such as fruit, nuts, latexes, resins, gums, medicinal plants, spices and dyes. The socio-economic importance of NTFP extraction to indigenous peoples is often underestimated. But as this market of NTFP demand increases these products can also be under threat from overexploitation. One example of NTFP exploitation is the collection of plants for the horticultural market. For certain groups of species, specialist collectors threaten the species by targeting individual rare species in the wild. This is the case, for example, with cacti and other succulents, orchids, cycads and trees such as magnolia and camellia. Collecting affects a relatively small set of species, but with devastating effects. The demand for wildlife products in general drives an illegal trade estimated to be worth up to US$10 billion per year (Haken, 2011), ranking it amongst the top international crimes. Orchids are popular plants in the legal horticultural trade but are also traded illegally (Hinsley et al., 2015). The global assessment of cacti (Goettsch et al., 2015), recently published by the IUCN SSC Cactus and Succulent Specialist Group, found that 31 per cent of cactus species are threatened with extinction. The illegal trade of live plants and seeds for the horticultural industry and private collections, as well as their unsustainable harvesting are the main threats to cacti, affecting 47 per cent of threatened species (Goettsch et al., 2015). This extensive traffic of wild-origin plants is a major threat to biological diversity creating a need for legal protection of traded species (Flores-Palacios and Valencia-Diaz, 2007).

Because the trade in wild plants, and animals, crosses national borders, the effort to regulate it requires international cooperation to safeguard desired species from overexploitation. The Convention on the

International Trade in Endangered Species (CITES) aims to ensure that international trade in specimens of wild animals and plants does not threaten their survival. The species covered by CITES are listed in three appendices.

- Appendix I includes species threatened with extinction. Trade in specimens of these species is permitted only in exceptional circumstances.
- Appendix II includes species not necessarily threatened with extinction, but in which trade must be controlled in order to avoid utilisation incompatible with their survival.
- Appendix III contains species that are protected in at least one country, which has asked other CITES parties for assistance in controlling the trade.

Actions to implement CITES effectively require knowledge of the distribution, abundance and conservation of species in the wild to ensure that levels of exploitation for international trade are sustainable. Very often this information remains limited, for example for heavily exploited species of *Dalbergia* referred to in Box 3.1.

Box 3.1 *Illegal Trade in Precious Timber*

In 2014 an assessment by UNEP concluded that the revenue from illegal logging and forest crime greatly exceeded other forms of illegal wildlife trade with an annual value of US$30–100 billion annually.

The scale of the problem is illustrated by two examples from the United Nations Office on Drugs and Crime's (UNODC) World Wildlife Crime Report 2016 focusing on the trade in rosewood and agarwood. Neither common name refers to a particular species, rosewood timber is derived from species of *Dalbergia* and *Pterocarpus* and, amongst other traditional uses, is used to make *hongmu* (meaning red wood) furniture and artworks in China. Dramatic growth in demand in recent years has led to the opening up of new illegal trade routes, with a particularly heavy impact on *Dalbergia* species in Madagascar. Three thousand tons of Malagasy rosewood was seized in Singapore in 2014 – possibly the largest ever seizure of illegal wildlife.

Agarwood (also known as *oud*), traditionally used in perfumes and incense, derives from the resinous wood produced in wounded trees of *Aquilaria* and a few other genera of Thymeleaceae. The UNODC

> report describes agarwood as 'likely the single most value-intensive wildlife commodity' with a kilogram of high quality agarwood chips being worth hundreds of thousands of dollars. Agarwood is cultivated in Bhutan, India, Malaysia, Myanmar, Vietnam and elsewhere but wild-origin material attracts significantly higher prices. The impact of illegal harvesting in Hong Kong, named 'Fragrant Harbour' for the trade in incense, has attracted particular attention. *The Economist* (2016) reported that a single gram of high quality agarwood was worth USD 1,600, more than gold.
>
> Various species of *Dalbergia*, *Pterocarpus* and all *Aquilaria* are included in the Appendices of CITES and various efforts are underway to tackle illegal trade. But the problems of implementation and enforcement are immense given the huge demand for the products and the relatively limited capacity of CITES Authorities in at least some of the source countries.
>
> Sources: Nellemann *et al.*, 2014; Ratsimbazafy *et al.*, 2016; UNODC, 2016; *The Economist*, 2016.

Overexploitation on a commercial scale is a clear threat but subsistence use of species, that may have occurred for centuries may also start to be unsustainable due to increased human population and decrease in the extent of habitat supporting the species.

3.3 Invasive Species, Pests and Diseases

An increase in human movement has led to increased spread of invasive species, including pests and diseases, forming another major threat to plant diversity. Prior to human-induced movement of species, dispersal events of these species were driven by stochastic and other rare but 'natural' events. However, with increased travel and trade, organisms (including invasive species, pests and disease vectors) are being moved further and more frequently than ever before.

The increasing globalisation of trade in plants and plant material has led to a rapid increase in the introduction and spread of new and damaging plant pests and pathogens. Past examples of the devastating impact these organisms can have on plant populations, such as Dutch elm disease on UK elm trees and the emerald ash borer on US ash populations, illustrate the significant threat these alien pests and pathogens pose to global plant health.

Introduced organisms can have large ecological and economic impacts on their new ranges and the native species which inhabit them. One reason for damage done by plant pests and pathogens is that plants, having not evolved alongside these introduced species, have not evolved a natural resistance. Similarly, it is unlikely that natural predators will be present to control population sizes.

A new initiative – the International Plant Sentinel Network (see Box 3.2) – is set up to serve as an early warning system for detecting new and emerging plant pests and pathogens. Identifying potential

---

Box 3.2 *The International Plant Sentinel Network (IPSN)*

The International Plant Sentinel Network (IPSN) provides an early warning system to recognise new and emerging pest and pathogen risks on a global scale. It does this via a network of both national and international partnerships between National Plant Protection Organisations (NPPOs), plant protection scientists and botanic gardens and arboreta around the world.

The IPSN provides an opportunity for botanic gardens to build on their work in research and conservation by helping to safeguard plants from pests and pathogens. Botanic gardens help to provide scientific evidence regarding known quarantine organisms and potential new risks to NPPOs in order to inform plant health activities.

The IPSN aims to:

- seek and share examples of best practice;
- develop standardised methodologies for monitoring and surveying of damaging plant pests and pathogens;
- provide training materials to increase capability among member gardens;
- facilitate access to diagnostic support;
- develop databases in order to share and store information; and
- communicate scientific evidence with NPPOs.

The IPSN currently (2016) has 29 member gardens in Europe, Asia, Oceania and Africa. In addition, it works closely with the Plant Sentinel Network which is centred in North America.

threats to native flora before an organism is actually introduced can drastically improve the chances of eradication or control, or even better prevent the introduction in the first place. This is where sentinel plants can play an important role.

Invasive species are those that can spread quickly to become common and dominant, posing a great threat to native species. Invasive species do not have to be 'alien' species (non-native), they could be a native species that under new conditions become invasive. Invasive non-native species are a major problem in many areas, especially on islands and freshwater habitats, which historically have evolved without the pressure of these species. There has been increased awareness of the importance to control the impact of invasive alien species, but implementation is challenging.

3.4 Climate Change

Climate change is a highly complex threat that impacts plants on a variety of levels (Hawkins et al., 2008). Climate change is predicted to increase carbon dioxide levels in the atmosphere, alter temperature and precipitation and lead to rising sea-levels. On one hand for some species increase in carbon dioxide levels may lead to an increase in growth rate, however, the added complexity in response to changes in temperature, precipitation and sea-level rise will mean that their response is difficult to predict.

Associated with climate change is the increasing occurrence of extreme events such as hurricanes, storms, tsunamis, high winds and flooding which may have devastating impacts on plant species and assemblages.

Understanding the effects of climate change on plant species and communities requires long-term datasets. Some datasets exist, such as long-term phenological records (see Box 3.3 for an example of a data collection scheme) for certain plant species. But the information is only available in some regions, which makes global analysis difficult. Also many of the long-term data collection protocols and species selection for phenological studies were not initially set up to answer questions about climate change.

Similarly, experimental approaches are expensive and lengthy, so research in this field relies heavily on modelling. Models are used for predicting responses on many different levels, including single species, multi-species assemblages and global vegetation patterns. Many models are often based on predicted bioclimatic layers (including temperature and precipitation) and known points of presence of plants to predict

> Box 3.3 *Project BudBurst*
>
> Project BudBurst is a national citizen science initiative that engages the public in observations of phenological (plant life cycle) events and raises awareness of climate change.
>
> The goals of Project BudBurst are to (1) increase awareness of phenology as an area of scientific study; (2) increase awareness of the impacts of changing climates on plants; and (3) increase science literacy by engaging participants in the scientific process.
>
> Project BudBurst was launched in 2008, and has since engaged participants of all ages and backgrounds to record the timing of leafing and flowering of wild and cultivated species in the US through its online educational and data-entry programme.
>
> Observations of the timing of leafing, flowering and fruiting phases of plants throughout the year are all valued and recorded. Scientists and educators then use the data to learn more about how plant species respond to changes in climate locally, regionally and nationally. Project BudBurst data are freely available for anyone to download.
>
> Citizen science programmes such as Project BudBurst provide the opportunity for anyone to actively participate in scientific research. Such programmes are important in providing opportunities for individuals to contribute to a better understanding of climate change.
>
> Project BudBurst is a Windows to the Universe Citizen Science programme managed by the University Corporation for Atmospheric Research, the Chicago Botanic Garden, University of Montana in collaboration with the USA National Phenology Network and with financial support from the US Bureau of Land Management, US Geological Survey, NEON and the Fish and Wildlife Foundation.
>
> More information on Project BudBurst can be found online at www.budburst.org.

bioclimatic envelopes of species, the geographic movement of these envelopes are then predicted into the future (or past). Models are only as good as the data and assumptions on which they are built and are continually improving as data is refined and tested.

A recent meta-analysis of predicted extinction risks from climate change shows that overall, 7.9 per cent of all species are predicted to become extinct from climate change (Urban, 2015). Assumptions about dispersal significantly affected extinction risks and extinction risks from climate change are expected not only to increase but to accelerate for every degree rise in global temperatures (Urban, 2015).

In addition, pressure from climatic change on the other threats of exploitation, habitat loss, invasives, pests and diseases, are likely to be exacerbated by climate change. Even species that are not directly at risk of extinction by climate change could experience substantial changes in abundance, distribution and species interactions, which in turn could affect ecosystem stability and the provision of ecosystem services (Urban, 2015).

3.5 Conservation Assessments

The major threats to biodiversity have an overall negative impact on plant species with threats impacting on individual species in different ways. The combinations of some of the drivers of change also have compound effects that are difficult to predict. Therefore, in order to make an accurate evaluation of the status of plants in the wild, conservation assessments on a species level are needed. Without species-level conservation assessments, we can make general predictions of the state of the habitat or ecosystem, we may also be able to predict what may happen to some of the dominant or charismatic species, but only by looking at species-level conservation assessments across the board are we able to get a larger picture of the status of plants on our planet. In addition, species conservation assessments are essential to prioritise conservation action and to make informed conservation decisions.

There are different systems available for assessing species' extinction risk, but the most commonly used and most widely recognised system is the IUCN Red List of Threatened Species (IUCN, 2016). The International Union for Conservation of Nature (IUCN) has produced red data lists for over 50 years. Other systems are also available, including NatureServe's G-ranks, which is widely used in North America (see Box 3.4). The IUCN Red List system allows for both global and national assessment with some modifications (IUCN, 2012), and many national red list initiatives are based on the IUCN Red List system.

The IUCN Red List Categories and Criteria (IUCN, 2001) were developed to measure the risk of extinction to a species. They aim to identify species at high risk of extinction in the near future. The IUCN Red List system does not list species that are 'naturally

> Box 3.4 *Plant Conservation Assessments in the US*
>
> In the US there are various systems for assessing the conservation of plants, the most comprehensive being provided by NatureServe. This non-profit organisation maintains an online database recording the status of biodiversity in the US and Canada. Each native species is assigned a conservation status rank from one to five at the global, national and state level. The conservation ranks are based on a series of criteria including size of distribution range, species population trends and threats. Currently (May 2016) about one-third (5,732 species) of the US native vascular plant flora is identified by NatureServe as extinct, possibly extinct or threatened in the wild.
>
> The US Endangered Species Act (ESA), which was signed into law in 1973, aims to protect species from extinction by prohibiting actions that threaten their survival. The species that are protected are listed as threatened or endangered following a lengthy process of review. A species must be listed under ESA if it is threatened or endangered due to any of the following:
>
> - present or threatened destruction, modification or curtailment of its habitat or range;
> - overutilisation of the species for commercial, recreational, scientific or educational purposes;
> - disease or predation;
> - inadequacy of existing regulatory mechanisms; and
> - other natural or manmade factors affecting its continued existence.
>
> 'Endangered' means a species is in danger of extinction throughout all or a significant portion of its range. 'Threatened' means a species is likely to become endangered within the foreseeable future.
>
> At present, 796 plant taxa are protected by ESA, a relatively small proportion of the threatened plants identified by NatureServe. Once a species is listed, the US Fish and Wildlife Service is required to create a recovery plan to improve management of the species and increase its numbers so that it has a greater chance of survival.

rare' but at no risk of extinction. It predicts the probability of extinction within a specific time period and aids, but does not directly set, conservation priorities (Mace *et al.*, 2008).

3.6 The IUCN Red List Categories and Criteria

The IUCN Red List categories and criteria are intended to be an easily and widely understood system for classifying species at high risk of global extinction. The IUCN Red List system consists of nine different categories: extinct (EX), extinct in the wild (EW), critically endangered (CR), endangered (EN), vulnerable (VU), near threatened (NT), least concern (LC), data deficient (DD) and not evaluated (NE) (see Figure 3.2). The three threatened categories are CR, EN and VU. Taxa that do not qualify for a threatened category, but are close to qualifying for, or are likely to qualify for a threatened category in the near future, can be assigned to the category NT. The category LC is used for species that are assessed but are not considered threatened, including widespread species and rare but stable species. The use of the category DD may be assigned to poorly known taxa. Species not yet evaluated are classified as NE.

The IUCN Red List uses five different criteria to measure symptoms of extinction risk. The criteria refer to fundamental biological processes underlying population decline and extinction. But given major

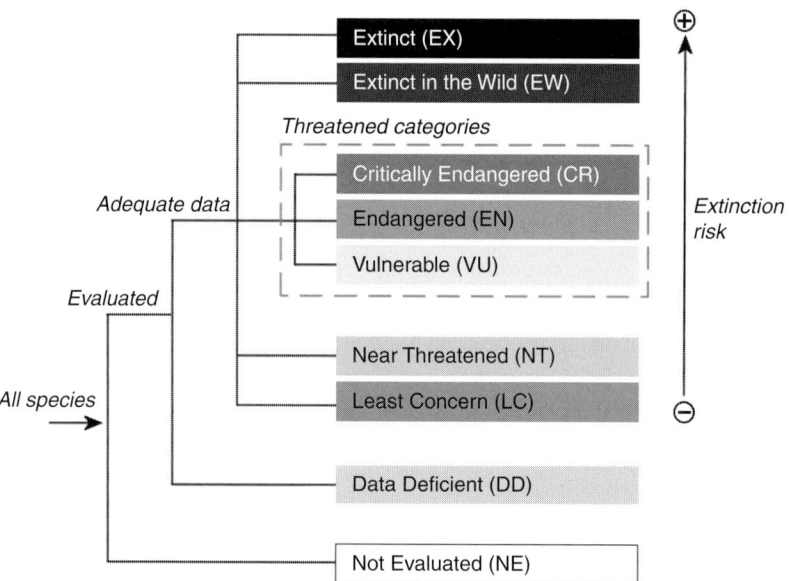

Figure 3.2 IUCN Red List categories of threat. Reused with permission from IUCN. For the colour version, please refer to the plate section. In some formats this figure will only appear in black and white.

differences between species, the threatening processes affecting them, and the lack of knowledge relating to most species, the IUCN system had to be both broad and flexible to be applicable to the majority of described species. In order to assess whether a species belongs to a threatened category (CR, EN, VU), species are evaluated in relation to five criteria: (1) population reduction; (2) geographic range; (3) small population size and decline; (4) very small or restricted population; and (5) quantitative analysis. The five criteria are based on a set of quantitative thresholds and several subcriteria.

Assessors are encouraged to evaluate taxa using all five criteria, but a taxon only needs to fulfil one of the five criteria to qualify for a threatened category. When several criteria are met, resulting in different status assessments, a precautionary but realistic attitude should be adopted (IUCN, 2001). Applying the highest category of threat (from reliable data) ensures a precautionary approach, rather than an evidentiary attitude, to making urgent decisions based on limited information (Collen et al., 2016).

The categories and criteria can be applied to any taxonomic level at or below the species level; however, the IUCN Red List requires an assessment of the full species before an assessment of infraspecific rank can be carried out (IUCN Standards and Petitions Subcommittee, 2016). It is also recommended that species are re-evaluated at least once every 10 years, as far as possible.

The criteria were designed to ensure consistency, transparency, and validity of its categorisation system, although some aspects will be interpreted differently for different taxonomic groups – therefore in addition to the categories and criteria, that have remained unchanged since 2001 – there are extensive guidelines available to facilitate the process for the assessors that help to interpret the criteria with data from different species groups. (IUCN Standards and Petitions Subcommittee, 2016).

For many plant red list assessments, a wide range of resources are consulted to gather all the required information. Sources include: national and regional floras, taxonomic databases, scientific papers, published and unpublished reports, expert knowledge, herbarium records and national red lists. Using all the available information in published and grey literature, a conservation category is assigned based on quantitative thresholds in the categories and criteria that need to be met by this data. All assessments are verified and reviewed by an expert reviewer before publication on the IUCN Red List. Once the conservation assessment is completed and reviewed, the assessments can be published. Both threatened and not threatened species are included on the IUCN Red List.

For more information, including free online training material, as well as general guidance please refer to the IUCN Red List website.[1]

The IUCN Red List was developed to suit all eukaryotic organisms, and has sometimes been criticised for being 'animal-focused'. The thresholds used in the threat analysis were set by experts specialising in many different taxonomic groups (Mace *et al.* 2008), and are used fairly evenly between different groups (Collen *et al.*, 2016). Nearly half of all threatened species with red list assessments on the IUCN Red List are of plants (11,643 of 24,304) (IUCN, 2016), this despite the fact that only 6 per cent of all plant species have been assessed globally and added to the IUCN Red List (Sharrock, *et al.*, 2014). Ninety-four per cent of all plants still remain unlisted.

IUCN Red List assessments have been comprehensively undertaken for some groups of organisms including: birds (BirdLife International, 2000, 2004), mammals (Baillie and Groombridge, 1996), amphibians (Stuart *et al.*, 2004; IUCN, Conservation International and NatureServe, 2008), as well as some plant groups such as cacti (Goettsch *et al.*, 2015), conifers (Farjon and Page, 1999; IUCN, 2016), cycads (Donaldson, 2003) and Magnoliaceae (Rivers *et al.*, 2016). As we have global red list assessments for so few plants (only 6 per cent), the question is whether we can estimate how many of the world's plants are at risk of extinction? Of the plant assessments on the IUCN Red List, 52 per cent are considered threatened (IUCN, 2016). However, the plants currently on the IUCN Red List are not a representative sample of plants overall as many of the groups assessed to date are biased towards species thought to be at risk. More recently there has been a move to also assess species that are not known to be at risk – and therefore the level of threatened plants on the IUCN Red List is actually declining over time (see Figure 3.3). In 1998, the threatened plant species represented 83 per cent of the species listed, whereas only 54 per cent of species listed in 2015 are considered threatened.

The Sample Red List Index for plants is an attempt to assess the level of threat for plants across different taxonomic groups. It aims to give a non-biased estimate of extinction risk by assessing a random selection of 1,500 species across five groups of plants (ferns, bryophytes, gymnosperms, monocotyledons and dicotyledons) (Brummitt *et al.*, 2015). Their result show that 22 per cent of plants are threatened in the wild. However, the estimates of threat within each of the five groups vary from 11 per cent

Conservation Assessments & Threats to Plant Diversity · 63

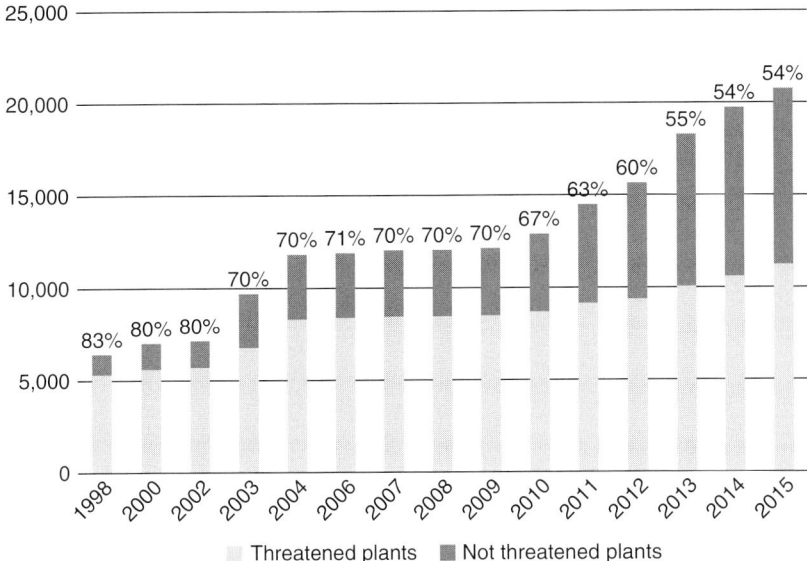

Figure 3.3 The number of plant species (threatened and not threatened) on the IUCN Red List over time, with the percentage of threatened plant species. Data from www.iucnredlist.org. For the colour version, please refer to the plate section. In some formats this figure will only appear in black and white

(dicotyledons) to 40 per cent (gymnosperms). The estimate of 22 per cent threat is calculated when each of these five groups is given equal weighting. But we know that the conifers are a very small group with just over 1,000 species, whereas there are over 200,000 species of dicotyledons. If instead each group was weighted relative to its number of species (as defined on *The Plant List*, 2013), the estimated level of threat across plants is much lower at 14 per cent.

Instead of taking a sampled approach, we can also look at some of the groups that have been assessed in their entirety – i.e. there has not been a bias in selecting the species but all species in a genus or family have been assessed. As mentioned previously very few groups of plants have been fully assessed, but those that have can give insight to the level of threat for plants. Even here we see a wide variation in the level of threats. Table 3.1 shows that for fully (or nearly) comprehensive groups of plants the level of threat varies from 17 per cent (Betulaceae) to 63 per cent (cycads). The first group is a plant family with many widespread temperate species, whereas cycads as a group consist of many range-restricted tropical species highly targeted for horticultural collections. The majority of

Table 3.1 *Examples of plant groups (genera, families, orders) of plants that have been comprehensively assessed using the IUCN Red List categories and criteria.*

Group of plants	Threatened species (%)	Reference
Cycads (Cycadales)	63	IUCN Red List (2016)
Magnoliaceae	48	Rivers et al. (2016)
Gymnosperms	40	IUCN Red List (2016); Brummitt et al. (2015)
Cactaceae	31	Goettsch et al. (2015)
Acer	28	Gibbs et al. (2009)
Quercus	27	Oldfield and Eastwood (2007)
Rhododendron	27	Gibbs et al. (2011)
Betulaceae	17	Shaw et al. (2014)

groups in Table 3.1 are trees, not surprisingly as tree conservation assessments have long been a specific priority of the IUCN SSC Global Tree Specialist Group and BGCI following the first attempt to globally assess trees with a report published in 1998 (Oldfield, et al. 1998). Since 2003, taxonomic and geographical groups of trees have been assessed systematically. Building on this important work, BGCI and the Global Tree Specialist Group have launched the Global Tree Assessment (Box 3.5) that will assess the conservation status of all tree species by 2020.

In addition to global, sometimes taxonomically focused, initiatives, there are many national red listing efforts that have assessed the national or regional conservation status of their plants. Some countries and regions have already fully red listed their entire flora, other countries are in the process of red listing their flora. Currently 96 countries have a national plant red list (National Red List, 2016). For a sample of national plant red listing efforts see Table 3.2. Currently some 20,755 plant species have global IUCN Red List assessments, but a further 90,000 plant names have national, regional or non-IUCN conservation assessments (data from National Red List and BGCI, unpublished). The latter represents assessments for 26 per cent of all plants respectively, progress towards, but still a long way from, assessments of all plant species.

National red list assessments are assessing the risk of extinctions of plants within national (or other defined regional) boundaries. National assessments reflect the risk of extinction of the species within the national boundaries but species may freely cross these boundaries, therefore the national or regional population may only be a subset of the total.

Box 3.5 *Global Tree Assessment*

The Global Tree Assessment aims to provide conservation assessments of all the world's tree species by 2020.

Despite the importance of trees, many are threatened by overexploitation and habitat destruction, as well as by pests, diseases, drought and their interaction with global climate change. In order to estimate the impact of such threats to trees there is an urgent need to conduct a complete assessment of the conservation status of the world's tree species – the Global Tree Assessment.

The assessment will identify those tree species that are at greatest risk of extinction. The goal of the Global Tree Assessment is to provide prioritisation information to ensure that conservation efforts are directed at the right species so that no tree species becomes extinct.

In order to achieve this target a complete global list of the world's tree species needs to be generated: GlobalTreeSearch. This list will help us to track progress towards the goal, and be used for gap analysis to set priorities both taxonomically and geographically for conservation assessment efforts.

The Global Tree Assessment is an initiative led by BGCI and the IUCN/SSC Global Tree Specialist Group. Work is ongoing to develop an even more extensive global collaborative partnership, involving the coordinated effort of many institutions and individuals. These steps will enable the Global Tree Assessment to achieve its 2020 target.

Assessments of endemic species (species found only within the country or region for which the red list was made) can be considered global assessments and simply follow the same categories and criteria as a global assessment. Assessments of non-endemic species have to take into account the possibility of immigration into the region and also whether the regional population may serve as an important source for population as a whole (IUCN, 2012). National conservation assessments of a species can have the same risk of extinction as the global population, or a species

Table 3.2 *Examples of national plant species conservation assessments*

Country	Nationally Threatened, %	Number of threatened plants	Species assessed (entire flora)	Source
Australia	?*	1,257	1,257 (20,000–30,000)	EPBC Act List of Threatened Flora[a]
Brazil	47.7*	2,479	5,195 (46,097)	CNCFlora[b]
China	10.6	3,767	35,610 (35,610)	Ministry of the Environmental Protection of the People's Republic of China[c]
Great Britain	19.6	345	1,756 (1,756)	Cheffings and Farrell (2005)[d]
Madagascar	56.3*	540–1,676	960 (11,350)	Catalogue of the Vascular Plants of Madagascar[e]
South Africa	13.4	2,657	20,295 (20,295)	SANBI (2015) Red List of South African Plants[f]
South Korea	41.2*	224	543 (3,304)	KORED[g]

Sweden	17.4	270	1,549 (1,549)	Artdatabanken[h]
USA (natives, species only)	32.5	5,732	17,640 (17,640)	NatureServe[i]

Notes: ⋆ Flora not fully assessed yet (May 2016)

[a] www.environment.gov.au/cgi-bin/sprat/public/publicthreatenedlist.pl?wanted=flora
[b] http://cncflora.jbrj.gov.br/
[c] www.zhb.gov.cn/gkml/hbb/bgg/201309/t20130912_260061.htm
[d] http://jncc.defra.gov.uk/page-3354
[e] www.tropicos.org/Project/MADA
[f] http://redlist.sanbi.org/
[g] www.korearedlist.go.kr/
[h] www.artdatabanken.se/en/the-red-list/
[i] www.natureserve.org/

may be at higher risk of going extinct (nationally/regionally) than the global assessment. In some cases, a species could even be less at risk nationally than globally, this could happen if there are no threats to the species within the regional boundaries under which it is assessed.

Regional and national red lists provide countries with key information about species status within their borders, which can be used directly for national conservation and planning policies supporting effective protection of biodiversity. The Convention on Biological Diversity (CBD) is implemented on a national scale, and countries report on their biodiversity on a national level, for which national red list assessments are important. National and regional assessments also help build capacity, expertise and conservation interest within a given region. The preparation of red list assessments at subglobal levels further enables far more information to be generated and fed into the global assessments (IUCN Red List Committee, 2013).

3.7 Influencing Policy and Legislation

Plant red list assessments determine the conservation status of plants in the wild. This information can then be used to prioritise plants most at risk of extinction and in need of conservation action. However, a red list conservation assessment is only one of the many steps in the prioritisation for species conservation. Conservation prioritisation should also take into account not only the extinction risk but also cost, chance of success and other factors (Collen *et al.*, 2016).

In addition to the importance of conservation assessments to guide effective conservation of biodiversity, conservation assessments are also used to inform policy and various conventions supporting effective protection of biodiversity. Species conservation assessments are being used to report and measure progress towards the UN's Sustainable Development Goals and the Aichi Biodiversity targets set by the Convention on Biological Diversity (CBD).

All global plant conservation assessments contribute towards the second target of the Global Strategy for Plant Conservation (GSPC):

GSPC Target 2: An assessment of the conservation status of all known plants, as far as possible, to guide conservation action.

In addition to global assessments contributing to the GSPC, national and regional red lists are essential resources to inform national biodiversity strategies and action plans (NBSAPs), and to establish a baseline measure of the status of biodiversity, from which the causes of biodiversity loss can be identified and conservation priorities can be established. Red list conservation assessments on a national scale also directly help countries monitor their progress towards the Aichi Biodiversity Targets, particularly target 12.

Aichi Target 12: By 2020 the extinction of known threatened species has been prevented and their conservation status, particularly of those most in decline, has been improved and sustained.

As currently a minority of plants have global conservation assessments, we urgently need to increase our knowledge of plants and the threat they face, in order for plants to be a key player in more effectively influencing policy and legislation on an international level.

3.8 Conclusion

Plant diversity is under ever increasing pressure from a range of different threats but we have limited knowledge of the global conservation status of the majority of plant species around the world. With increasing impacts of climate change and other threats, species-level conservation assessments are urgently required to plan more effective *in situ* and *ex situ* conservation strategies with the limited resources available.

There is evidence that extinction is altering key ecosystem processes and further species loss will accelerate these changes; the impact of species loss is now comparative to other global environmental change in altering ecosystem function (Hooper *et al.*, 2012). Therefore, if this loss of plant biodiversity is not stemmed, countless opportunities to develop new solutions to pressing economic, social, health and industrial problems will also be lost.

Note

1. See www.iucnredlist.org.

References

Baillie, J. and Groombridge, B. (Eds) (1996). *1996 IUCN Red List of Threatened Animals*, IUCN, Gland, Switzerland.

BirdLife International (2000). *Threatened Birds of the World*. Barcelona and Cambridge, UK: Lynx Edicions and BirdLife International.

BirdLife International (2004). *Threatened Birds of the World 2004*, CDROM BirdLife International, available online at www.birdlife.org/datazone/species/index.html [accessed February 2017].

Brummitt, N. and Bachman, S. (2010). *Plants Under Pressure: A Global Assessment: The First Report of the IUCN Sampled Red List Index for Plants*. Richmond, UK: Royal Botanic Gardens, Kew.

Brummitt, N. A., Bachman, S. P., Griffiths-Lee, J. et al. (2015). Green plants in the red: a baseline global assessment for the IUCN Sampled Red List Index for Plants. *PloS ONE*, 10(8), e0135152.

Cheffings, C. M. and Farrell, L. (Eds) (2005). *The Vascular Plant Red List for Great Britain*. Peterborough, UK: JNCC.

Collen, B., Dulvy, N. K., Gaston, K. J. et al. (2016). Clarifying misconceptions of extinction risk assessment with the IUCN Red List. *Biology Letters*, 12(4), 20150843.

Corlett, R. T. (2016). Plant diversity in a changing world: status, trends and conservation needs. *Plant Diversity*, 38(1): 11–18.

De Vos, J. M., Joppa, L. N., Gittleman, J. L., Stephens, P. R. and Pimm, S. L. (2015). Estimating the normal background rate of species extinction. *Conservation Biology*, 29(2): 452–462.

Donaldson, J. (Ed.) (2003). *Cycads: Status Survey and Conservation Action Plan*, IUCN-SSC Cycad Specialist Group, available online at https://portals.iucn.org/library/sites/library/files/documents/2003-010.pdf [accessed February 2017].

Economist, The, (2016) Fragrant Arbour: Thieves are destroying the tree that gave Hong Kong its name, 20 February.

Farjon, A. and Page, C. N. (Eds) (1999). *Conifers: Status Survey and Conservation Action Plan*, IUCN-SSC Conifer Specialist Group, available online at https://portals.iucn.org/library/sites/library/files/documents/1999-024.pdf [accessed February 2017].

Flores-Palacios, A. and Valencia-Diaz, S. (2007). Local illegal trade reveals unknown diversity and involves a high species richness of wild vascular epiphytes. *Biological Conservation*, 136(3): 372–387.

Gibbs, D., Gibbs, Y. and Chen, Y. (2009). *The Red List of Maples*. Richmond, UK: Botanic Gardens Conservation International.

Gibbs, D., Chamberlain, D. and Argent, G. (2011). *The Red List of Rhododendrons*. Richmond, UK: Botanic Gardens Conservation International.

Goettsch, B., Hilton-Taylor, C., Cruz-Piñón, G. et al. (2015). High proportion of cactus species threatened with extinction. *Nature Plants*, 1: 15142.

Haken, J. (2011). *Transnational Crime in the Developing World*. Washington, DC: Global Financial Integrity, available online at http://www.gfintegrity.org/storage/gfip/documents/reports/transcrime/gfi_transnational_crime_web.pdf [accessed May 2016].

Hawkins, B., Sharrock, S. and Havens, K. (2008). *Plants and Climate Change: Which Future?* Richmond, UK: Botanic Gardens Conservation International.

Hinsley, A., Verissimo, D. and Roberts, D. L. (2015). Heterogeneity in consumer preferences for orchids in international trade and the potential for the use of market research methods to study demand for wildlife. *Biological Conservation*, 190: 80–86.

Hooper, D. U., Adair, E. C., Cardinale, B. J. *et al.* (2012). A global synthesis reveals biodiversity loss as a major driver of ecosystem change. *Nature*, 486(7401): 105–108.

IUCN (2001). *IUCN Red List Categories and Criteria: Version 3.1.* Prepared by the IUCN Species Survival Commission, available online at http://s3.amazonaws.com/iucnredlist-newcms/staging/public/attachments/3097/redlist_cats_crit_en.pdf [accessed March 2016].

IUCN (2012). *Guidelines for Application of IUCN Red List Criteria at Regional and National Levels:* Version 4.0. Gland, Switzerland and Cambridge, UK: IUCN, available online at https://portals.iucn.org/library/sites/library/files/documents/RL-2012-002.pdf [accessed March 2016].

IUCN (2016). *The IUCN Red List of Threatened Species.* Available online at www.iucnredlist.org [accessed May 2016].

IUCN Red List Committee (2013). *The IUCN Red List of Threatened Species™ Strategic Plan 2013–2020.* Version 1.0. Prepared by the IUCN Red List Committee.

IUCN Standards and Petitions Subcommittee. (2016). *Guidelines for Using the IUCN Red List Categories and Criteria. Version 12.* Prepared by the Standards and Petitions Subcommittee. Available online at http://www.iucnredlist.org/documents/RedListGuidelines.pdf [accessed May 2016].

IUCN, Conservation International, and NatureServe (2008). *An Analysis of Amphibians on the 2008 IUCN Red List.* Available online at www.iucnredlist.org/amphibians [accessed May 2016].

Joppa, L. N., O'Connor, B., Visconti, P. *et al.* (2016). Filling in biodiversity threat gaps. *Science*, 352(6284): 416–418.

Kettle, C. J. and Koh, L. P. (Eds) (2014). *Global Forest Fragmentation.* Wallingford, UK: CABI.

Mace, G. M., Collar, N. J., Gaston, K. J., Hilton-Taylor, C., Akçakaya, H. R., Leader-Williams, N., Milner-Gulland, E. J. and Stuart, S. N. (2008). Quantification of extinction risk: IUCN's system for classifying threatened species. *Conservation Biology*, 22(6): 1424–1442.

Millennium Ecosystem Assessment (2005). *Ecosystems and Human Well-Being: Synthesis.* Washington, DC: Island Press.

Myers, N., Mittermeier, R. A., Mittermeier, C. G., Da Fonseca, G. A. and Kent, J. (2000). Biodiversity hotspots for conservation priorities. *Nature*, 403(6772): 853–858.

National Red List (2016). National Red List. Available online at www.nationalredlist.org [accessed May 2016].

Nelleman, C., Henriksen, R., Baxter, P., Ash, N. and Mrema, E. (2014). *The Environmental Crime Crisis Threats to Sustainable Development from Illegal Exploitation and Trade in Wildlife and Forest Resources. A UNEP Rapid Response Assessment.* Nairobi and Arendal: UNEP and GRID Arendal.

Oldfield, S. and Eastwood, A. (2007). *The Red List of Oaks.* Fauna & Flora International, UK.

Oldfield, S. F., Lusty, C. and MacKinven, A. (1998) *The World List of Threatened Trees.* Cambridge, UK: World Conservation Press.

Ratsimbazafy, C., Newton, D. J. and Stéphane, R. (2016). *Timber Island: Rosewood and Ebony Trade of Madagascar.* Cambridge, UK: TRAFFIC.

Reynolds, J. D. and Peres, C. A. (2006). Overexploitation. In: Groom, M. J., Meffe, G. K. and Carroll, C. R. (Eds), *Principles of Conservation Biology.* Sunderland, MA: Sinauer, pp. 253–291.

Rivers, M., Beech, E., Murphy, L. and Oldfield, S. (2016). *The Red List of Magnoliaceae: Revised and Extended.* Richmond, UK: Botanic Garden Conservation International.

SANBI (2015). *Statistics: Red List of South African Plants*, version 2015.1. Available online at Redlist.sanbi.org [accessed 16 April 2016].

Sharrock, S., Oldfield, S. and Wilson, O. (2014). *Plant Conservation Report 2014: A Review of Progress in Implementation of the Global Strategy for Plant Conservation 2011–2020.* Montréal, Canada and Richmond, UK: Secretariat of the Convention on Biological Diversity, and Botanic Gardens Conservation International, Technical Series No. 81, 56 pp.

Shaw, K., Stritch, L., Rivers, M., Roy, S., Wilson, B. and Govaerts, R. (2014). *The Red List of Betulaceae.* Richmond, UK: Botanic Garden Conservation International.

Stuart, S. N., Chanson, J. S., Cox, N. A., Young, B. E., Rodrigues, A. S., Fischman, D. L. and Waller, R. W. (2004). Status and trends of amphibian declines and extinctions worldwide. *Science*, 306(5702): 1783–1786.

The Plant List (2013). A working list of all known plant species, version 1.1. Available online at www.theplantlist.org/ [accessed May 2016].

United Nations Office on Drugs and Crime (UNODC) (2016). *World Wildlife Crime Report: Trafficking in Protected Species.* New York, NY: United Nations.

Urban, M. C. (2015). Accelerating extinction risk from climate change. *Science*, 348 (6234): 571–573.

4 · *The Role of Botanic Gardens in* In Situ *Conservation*

JIN CHEN, RICHARD T. CORLETT AND CHARLES H. CANNON

4.1 Introduction

Biodiversity is both the heritage of four billion years' evolution on this planet and the fundamental basis for the future security and sustainability of humankind. Biodiversity includes the diversity of genes, species and ecosystems, as well as the interactions among species and the capacity for evolutionary changes and adaptation. As a result of accelerating anthropogenic impacts, including land-use change, overexploitation and pollution, the loss of biodiversity has been recognised as one of the main threats to the planet (Butchart *et al.*, 2010) and the insidious trends of climate change are predicted to result in a significant speeding up of species extinction (Urban, 2015). Measures for saving biodiversity include conservation *in situ* and *ex situ*, reinforcement and reintroduction, conservation translocation and assisted migration, direct ecological interventions and, potentially for some species, de-extinction. Botanic gardens have an actual or potential role in all these activities, and thus occupy a core position in the future of biodiversity, particularly for plants (Wyse Jackson and Kennedy, 2009; Heywood, 2011).

While botanic gardens are widely recognised as centres for the *ex situ* conservation of plants, in seed banks and living collections, their role in *in situ* conservation – conservation in natural and semi-natural environments – is frequently overlooked. Particularly in the tropics, however, where biodiversity-rich areas often have relatively low local conservation capacity (Chen *et al.*, 2009; Mammides *et al.*, 2016), botanic gardens can play a crucial role in *in situ* conservation and their involvement in this should be taken as imperative (Chen *et al.*, 2009). Botanic gardens located in developed countries are also often involved in local conservation efforts, and can provide important support for activities in biodiversity-rich but capacity-poor locations through partnerships with local botanic gardens and other institutions, and through involvement in capacity building.

These local activities by botanic gardens need to be placed in a broader, global context. There is a growing realisation among environmental scientists that the Earth has passed a major tipping point into a new geological epoch, co-dominated by human and natural influences, the Anthropocene (Williams *et al.*, 2016). This change reflects the evolution and global expansion of modern humans over the last 100,000 years, the origin and spread of agriculture over the last 10,000 years, increasing industrialisation over the last 200 years, and, most dramatically, the social, political and technological developments since 1950 known as the 'great acceleration' (Corlett, 2015). There is no going back. The Anthropocene thus poses a particular challenge to conservation, which has traditionally focused on keeping things the way they are or restoring them to the way they were (Corlett, 2016a). If neither of these is possible and the only way is forward into an uncertain future, what should conservation – and conservationists – be doing? The answer is not yet obvious, but any discussion of the role of botanic gardens in conservation must take these uncertainties into account.

4.2 An Expanded Concept of *In Situ* Conservation

Both *in situ* and *ex situ* conservation are important parts of an integrative approach to conservation, and each has certain advantages and disadvantages (see Table 4.1). *In situ* conservation – on-site conservation – is undoubtedly the fundamental measure for global biodiversity conservation (SCBD, 2010). It can conserve a great variety of threatened ecosystems, species and genes, including undescribed species and unrecognised genetic diversity. Furthermore, *in situ* conservation can allow the continued evolution and co-evolution of biological systems. In relation to botanic gardens in particular, it is also important to note that *in situ* conservation automatically includes both plants and animals, and their interactions, while *ex situ* conservation in botanic gardens usually involves only plants. In the last few decades, a significant achievement for *in situ* conservation worldwide has been reached mainly by setting up protected areas of various types, which now include an estimated 15 per cent of the land surface of the Earth (Sharrock *et al.*, 2014).

Table 4.1 *Advantages and disadvantages of* in situ *vs.* ex situ *conservation*

Items	*In situ* conservation	*Ex situ* conservation
Species α diversity protected	Varied depending on habitats	Often high
Protects plants, animals or both together	Plants and animals together	Plants and animals in separate facilities
Genetic integrity protected	Yes, if populations sizes adequate	Needs to be specifically targeted
Vulnerable to climate change	Often yes, particularly if small or isolated	Less so
Input requirements	Often very high	Relatively low for plants
Relevant to education and research	Depends on the situation	Often yes
Interest conflict with local community	Often yes, although many exceptions (Secretariat CBD, 2008)	Often not the case
Monitoring success	Often difficult	Relatively easy
Protect mutualists, parasites and other associated species	Yes	No, unless deliberately targeted
Allow continued adaptive evolution	Yes	No

The traditional practice and policy of *in situ* conservation is focused on wild populations in natural and semi-natural areas (see Box 4.1), but the advent of the Anthropocene means that all ecological communities are impacted by human activities to some degree and the concept of a 'wilderness' existing beyond our reach is extinct. Even old-growth forest communities are affected in many ways, from defaunation (Harrison *et al.*, 2013; Dirzo *et al.*, 2014), altered nutrient cycles (Holtgrieve *et al.*, 2011), to simple fragmentation, isolation and restricted gene flow (Haddad *et al.*, 2015). The scope of *in situ* conservation therefore needs to be expanded beyond protected 'natural' areas, like national parks and forest reserves, to include the full spectrum of areas where wild species grow spontaneously within their native ranges. Indeed, we should go further and include planted populations of native species within urban areas as these populations have the potential to act as breeding stock.

Box 4.1 *Summary of IUCN Protected Area Categories*

Category Ia (strict nature reserve): set aside to protect biodiversity, where human visitation, use and impacts are strictly controlled and limited to ensure protection of the conservation values.

Category Ib (wilderness area): usually large unmodified or slightly modified areas, retaining their natural character and influence, without permanent or significant human habitation, protected and managed to preserve their natural condition.

Category II (national park): protect large-scale ecological processes, along with the complement of species and ecosystems characteristic of the area, which also provide a foundation for environmentally and culturally compatible spiritual, scientific, educational, recreational and visitor opportunities.

Category III (natural monument or feature): protect a specific natural monument, which can be a landform, sea mount, submarine cavern, geological feature such as a cave or even a living feature, such as an ancient grove.

Category IV (habitat/species management area): protect particular species or habitats, where management reflects this priority. Many will need regular active interventions to address the requirements of particular species or to maintain habitats, but this is not a requirement of the category.

Category V (protected landscape): protected areas where the interaction of people and nature over time has produced an area of distinct character with significant ecological, biological, cultural and scenic value and where safeguarding the integrity of this interaction is vital to protecting and sustaining the area and its associated values.

Category VI (protected areas with sustainable use of natural resources): protects ecosystems and habitats, and associated cultural values and traditional natural resource management systems. Generally large areas, with most of the area in a natural condition, where a proportion is under sustainable natural resource management with low-level non-industrial use of natural resources compatible with nature conservation.

The Role of Botanic Gardens in *In Situ* Conservation · 77

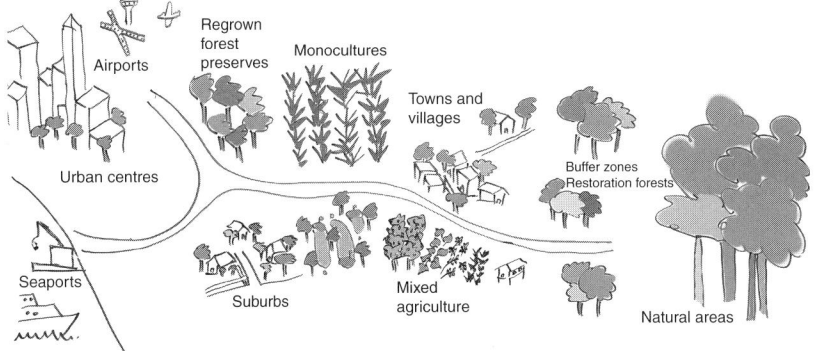

Figure 4.1 Opportunities and challenges for *in situ* conservation in the Anthropocene.

To fully integrate *in situ* conservation practice across the entire landscape, the transition between 'natural' and 'domesticated' landscapes should be seen as a continuous spectrum and not be defined in black and white terms. Across this continuum from pristine to urban vegetation, a range of opportunities exist for promoting positive *in situ* conservation practices (see Figure 4.1). We cannot focus on just one element in this dynamic and highly segmented landscape. For example, an increasing amount of evidence indicates that urban environments can be surprisingly diverse (Baldock *et al.*, 2015; MacGregor-Fors *et al.*, 2016), despite the fact little concerted effort has been made to promote diversity in these environments. Cities present a complex mixture of habitats, spatially chaotic, widely variable on a micro-climatic scale, with rapidly shifting land-use change. More cities are investing resources and research into their urban forests (Bernstein *et al.*, 2016; Pinto *et al.*, 2016), recognising the great benefit these trees provide to their citizens. Suburban areas can provide a considerable diversity of environments, often supporting smaller vertebrates (Soulsbury and White, 2016). In contrast to the ecological deserts of industrial agriculture and monoculture plantations that increasingly dominate 'rural' areas, suburban landscapes provide opportunities for the promotion of diverse plantings and the active involvement of local citizens.

The continuum of land uses from the urban centre to the heart of the forest provides a number of locations where both spontaneous regrowth and planted 'wild' species could used to promote *in situ* conservation.

4.3 What Can Botanic Gardens Offer to *In Situ* Conservation?

The value of setting aside large areas of land for biodiversity conservation is globally acknowledged. However, the overall effectiveness of the current global protected area system for conservation remains questionable. Management effectiveness varies greatly and around 50 per cent of the protected areas in tropics can be categorised as 'degraded' to some extent (Laurance *et al.*, 2012). The reasons for this failure are complex, but include underfunding, lack of management capacity, lack of monitoring and weak policy enforcement (Leverington *et al.*, 2010). Botanic gardens can contribute in various ways to enhancing the effectiveness of protected areas, and in some of these they can play an irreplaceable role.

First, protected areas cannot sustain biodiversity in a hostile surrounding environment. As Folke *et al.* (1996) pointed out, 'nature reserves and other protected areas are an important short-term step. Conservation efforts should be planned at the scale of the regional landscape. Small reserves will lose their distinctive species if they are surrounded by a hostile landscape.' The traditional expertise of botanic gardens in horticulture and crop systems, particularly in the tropics (Ali and Trivedi, 2011), can make a major contribution to the development of sustainable land-management practices at the landscape level, so that protected areas are buffered and connected as much as practical.

Second, the existing protected area network is far from ideal (Margules and Pressey, 2000; Le Saout *et al.*, 2013), especially under the scenario of climate change (Wang *et al.*, 2016). For example, a study of China's protected areas that combined high resolution climatic data with species distribution modelling indicated that, under the most extreme climate change scenario, about 16 per cent of the protected areas (totalling 233 protected areas in this study) are projected to lose 30 per cent of their current ranges (Y. B. Zhang *et al.*, 2014). In another study, in Yunnan, a biodiversity hotspot in Southwest China, a majority (1,400 species among 1,996 species studied) of the species are projected to lose more than 30 per cent of their current ranges (M. G. Zhang *et al.*, 2014). There is an urgent need to revise existing protected area networks, on both the regional (Le Saout *et al.*, 2013) and local scale, in order to allow species and ecosystems to adjust to climate change. These revisions need to incorporate knowledge from climate change modelling, phenological observations and ecophysiological studies. Botanic gardens often hold long-term observations on particular plant species (Primack and Miller-Rushing, 2009) and herbarium records can extend this timeline back further into the past. Some botanic gardens also conduct research on plant physiology and on plant and ecosystem responses to climate change, which can contribute to the revision of the protected area network.

Third, the small, isolated fragments of forest remaining along the margins of highways, between agricultural fields, and near villages and towns, can play a significant role in gene flow and continuity among larger patches of forests. Even though the plants in these fragments are certainly more vulnerable to increased mortality and poor health, botanic gardens could play a role in mapping, surveying, and categorising these fragments and understanding the major parameters of their management, who holds land tenure, what vulnerabilities exist, and prioritising among these fragments in their conservation value and probable sustainability and whether any type of ecological intervention is needed, such as eradication of invasive woody species. While the entire landscape needs to be considered, each site, from the city centre to the heart of the forest, will likely and unfortunately require specific policies that emanate from different governmental bodies and jurisdictions.

Finally, botanic gardens' expertise in *ex situ* conservation of plants can have direct applications in *in situ* conservation, through the ecological restoration of degraded sites, often using mostly common species, and the reinforcement of declining plant populations of rare species and the reintroduction of locally extinct species. Indeed, botanic gardens should see *ex situ* and *in situ* conservation as part of the same process, and manage *ex situ* and *in situ* populations of threatened species as part of a whole. For example, genetic material held *ex situ* can be used to enhance the genetic diversity of endangered wild populations, contributing to their long-term survival (Fotinos *et al.* 2015). However, the world is still a long way from achieving Target 8 of the GSPC, which is the *ex situ* conservation of at least 75 per cent of threatened plant species by 2020, with at least 20 per cent available for recovery and restoration programmes, with the gap between the target and reality being greatest in the hyperdiverse tropics (Corlett, 2016b).

4.4 How Are Botanic Gardens Best Involved in *In Situ* Conservation?

The opportunities for botanic garden involvement in *in situ* conservation are very diverse and the appropriate roles will vary between gardens, depending on their locations and expertise. Most botanic gardens, however, share a number of characteristics that set them apart from other institutions and have a particular relevance to conservation. First, they are neither purely academic nor purely anything else: they work at the interface between science, education and practice, which is a similar space to that occupied by real-world conservation. Second, they are, of necessity, part of their own local and regional communities, and have generally excellent relations with both government agencies and non-governmental groups, which can help them to influence decisions that affect conservation. Third, botanic gardens take a long view: a perspective that is equally important in garden management and in conservation. Finally, most botanic gardens have some degree of commitment to scientific research, with a frequent focus on systematics and taxonomy, which is highly relevant to conservation.

In many cases, botanic gardens themselves include patches of natural or semi-natural habitats within their area. Among the 2679 listed in the Botanic Garden Conservation International (BGCI) database, there are 409 botanic gardens (15.3%) which have natural habitats within them. The reported mean area of 164.5 ± 856.1 hectares (mean ± SD, $n = 286$) suggests that most of these habitat patches are too small to play a major role in conservation, but they provide experience in habitat management and a convenient location for research. In Xishuangbanna Tropical Botanic garden (XTBG), forest areas within the gardens have been used for research into orchid reintroduction techniques. In XTBG and other botanic gardens, these habitat patches are also major sites for environmental education (see Chapter 8). An example of a botanic garden undertaking *ex situ* conservation as an extension of its grounds is given in Box 4.2.

Box 4.2 Ex Situ *and* In Situ *Plant Conservation: An Example of Integrated Approach in Hawaii*

Limahuli Garden and Preserve is part of the National Tropical Botanic Garden (NTBG) and combines both a botanic garden and adjacent natural areas on the north shore of the Hawaiian Island of Kauaʻi. Preserves are a resource for the NTBG staff to develop and test conservation protocols in a range of ecosystems beyond those that exist in the gardens. Each of NTBG's preserves has different characteristics, and some have proved more successful than others.

The important point is that the work done in each preserve has been productive in terms of the experience provided and the lessons learned.

Limahuli Preserve (acquired in 1994) has two distinct management areas. The Lower Limahuli Preserve contains approximately 242 hectares of land that is closed to the public. The unspoiled Limahuli Stream has never been highly degraded by human impact, although the ancient Hawaiians used its waters to irrigate taro patches. Today, the stream has a diversity of native fauna, including Hawaii endemic fish and crustaceans. Over the past 100 years the Lower Preserve has seen a major decline in the population of native plants, primarily due to the introduction of feral cattle in the late 1800s. The loss of native species has been accompanied by the establishment of many alien species of plants. Ten years ago the Limahuli Garden staff began a plant-community restoration programme.

The Upper Limahuli Preserve covers approximately 162 hectares of land above Limahuli Falls and extends from about 1,600 feet at the top of the falls to 3,330 feet at the summit of Hono O Napali. At upper elevations, the vegetation is characteristic of montane rainforest, while at lower elevations it is characteristic of lowland rainforest. The Upper Preserve is remote, requiring the use of a helicopter to gain access. It was never intensely cultivated or modified by the ancient Hawaiians and was still considered to be a pristine ecosystem with very few non-native species until the early 1980s. In 1982, and again in 1992, the area was severely damaged by two powerful hurricanes which denuded the vegetation and spread aerial-borne alien weed seeds. In the past 20 years the area has also been subject to increased pressure from expanding populations of feral pigs. Since 1992, management activities in this remote area are focusing on control of the worst of the invasive plant species and control of the feral pigs.

Source: http://ntbg.org/resources/preserves.php#limahuli

Although there seem to be no relevant figures available on this, some botanic gardens also manage natural areas outside the garden itself. An exceptional example of this is the protected area established at Dinghu Shan by the South China Botanic Garden (see Box 4.3), which was the first protected area in modern China. This can serve as a strong symbol of the key role of botanic gardens in *in situ* conservation.

> **Box 4.3** *South China Botanic Garden (SCBG) Set Up the First National Protected Area in China: Dinghu Mountain and Lake Natural Reserve*
>
> Dinghu Mountain and Lake (Chinese: 鼎湖山) is located in Dinghu District, 18 km to the east of Zhaoqing City, in the Dayunwu Mountain Range, in Guangdong Province of southern China, with an area of 1,133 ha. Known as the 'green gem on the Tropic of Cancer', the mountain's peaks rise above ancient trees and waterfalls, providing fresh air and habitats rich in birds and other species. Since ancient times, this area has been a tourist attraction and a Buddhist sacred place.
>
> In 1956, several distinguished scholars (Bing Zhi, Qian Chongsu, Yang Weiyi, Qin Renchang, and Chen Huanyong, etc.) raised a proposal to the third session of the first National People's Congress of China asking to set up protected areas for preserving natural vegetation in different provinces for the purpose of research. The Dinghu Mountain and Lake Natural Reserve was set up as the first reserve. Since 1974, the reserve has been affiliated to the South China Botanic Garden, Chinese Academy of Sciences (previously the Guangdong Institute of Botany, during 1974–1979). On 17 December 1979, the reserve was recognised as one of the first three protected areas joining the Man and the Biosphere programmeme of UNESCO. A long-term ecological research station has been set up by SCBG within the reserve, and thriving research programmes on long-term forest dynamics, forest restoration and climate change are conducted.
>
> Source: http://dhf.cern.ac.cn/ (in Chinese)

This extended concept of *in situ* conservation to include urban areas like golf courses, larger greenspaces, and even cemeteries could provide further opportunities for managing communities and populations of spontaneous vegetation.

4.5 Surveys and Inventories

In situ conservation must start with knowing what species are present and which ones need protection. Therefore species inventories and conservation status assessments are a critical part of any *in situ* conservation programme. Without this it is not possible to assign conservation priorities to species and areas, or to assess the effectiveness of conservation management. Inventories are needed for both existing protected areas and potential future additions to the system, in order to assess gaps and try to fill them.

In-house expertise in botanic gardens usually focuses on plants, but broader coverage of the biota is possible in collaboration with others. In any case, plant surveys are more difficult and time consuming than surveys of animals since, unlike most of the latter, plants do not sing, walk past camera traps, or come to baits. Botanic gardens have traditionally been centres for taxonomy and plant inventories (Vovides *et al.* 2013), and many have made significant achievements in documenting regional floras in the past, including the colonial herbaria in Europe for many tropical floras, the Missouri Botanical Garden for the flora of Madagascar, and botanic gardens in Singapore, Malaysia and Indonesia for Southeast Asia, and the South African National Biodiversity Institute (SANBI) for documenting the plants and conservation status in South Africa (see Box 4.4). This capacity, although in some cases it needs to be re-strengthened, could be used in *in situ* conservation.

The first target of the GSPC for 2020 is 'an online flora of all known plants', but it is important to note that perhaps 10–20 per cent of all land plant species are still unknown to science (Pimm and Joppa, 2015) so the task will not be complete even when the first online version becomes available. A recent review estimated that, although 369,000 species of flowering plants are known to science, in the past 10 years new species have continued to be described at the rate of about 3,000 a year, with no sign of this levelling off (RBG Kew, 2016). The new discoveries include some spectacular species, such as a 45 m tall legume tree, *Gilbertiodendron maximum*, from Gabon which was recently described (van der Burgt, 2015). There is no data available on how many of these new species were described from botanic gardens, or in collaboration with them, but the percentage is probably high, since most large botanic gardens are involved in this work.

Moreover, it is one thing knowing that a species exists, but for conservation we need to know where it is (and where it isn't), and whether its conservation status merits special attention (Corlett, 2016b). The GSPC target 2 is an assessment of the conservation status of all known plant species by 2020, but we are currently very far from achieving this. Fewer than 20,000 plant species have had a global IUCN Red List assessment, and only a few groups of plants (including conifers and cycads) have been fully assessed, far fewer than for vertebrate taxa (Krupnick, 2013; see also Chapter 3). Many more species have been

> **Box 4.4** *Survey for* In Situ *Plant Conservation in South Africa*
>
> South Africa currently has 2,576 nationally threatened plant species many of which are endemic to the country. Since 2005, the South African National Biodiversity Institute (SANBI) Threatened Species Programme (TSP) has collected distribution data for threatened species. Over 57,000 herbarium records have been digitised and georeferenced. In addition, a volunteer network has been monitoring populations of threatened plants in their habitats as part of the Custodians of Rare and Endangered Wildflowers (CREW) programme (www.sanbi.org). Additional threatened plant data sources, from national and provincial conservation authorities, regional herbaria, and atlas and citizen science programmes, have been included. Spatial data are now available for 2,345 of the threatened species. Of these, 1,554 (66%) species have at least one record within a formally protected area (von Staden, unpublished data).
>
> Since 2005, South Africa has expanded protection of terrestrial ecosystems through the establishment of biodiversity stewardship programmes. Twenty-four contractual protected areas have been declared on private or communal land. Landowners retain title to the land and are recognised as the management authority of the protected area. Biodiversity stewardship is proving to be a highly cost-effective way to expand the protected area network for plant species with over 75,000 ha of additional land protected and a further approximately 360,000 ha of additional contract protected areas in development.
> Source: SANBI's Plant Conservation Strategy (available online at http://biodiversityadvisor.sanbi.org/planning-and-assessment/plant-conservation-strategy/)

assessed at a national or regional level, using IUCN Red List criteria, but these assessments tend to be less robust and are also incomplete. The GSPC target 5 is the protection 'with effective management in place for conserving plants and their genetic diversity' of at least 75 per cent of the most important areas for plant diversity of each ecological region by 2010. New initiatives for identifying these important plant areas (IPAs; Marignani and Blasi, 2012) have identified 1,771 globally so far, but very few are well protected currently and even fewer are well managed (Royal Botanic Gardens, Kew, 2016).

4.6 Monitoring of Biodiversity *In Situ*

Effective long-term *in situ* conservation requires species-level monitoring to ensure that viable populations of target species persist within protected areas. Species can be lost even when the habitat remains intact, but small, fragmented and disturbed protected areas are particularly likely to lose species, and ongoing climate change makes regular monitoring important everywhere. Despite its importance, however, biodiversity monitoring of protected areas is rare, particularly in the tropics. For example, in China a significant achievement has been reached, with 17 per cent of China's land in protected areas. However, although these protected areas have been generally successful in preserving habitats, their conservation effectiveness for targeted plants and animals is little known (Ren et al., 2015). A recent review of the effectiveness of tropical reserves (Laurance et al., 2012) had to rely on qualitative interviews rather than actual survey data, and they still found that half of protected areas were experiencing an erosion of biodiversity.

The species-level monitoring that currently exists is largely on an *ad hoc* basis. A notable exception to this is the 'The Center for Tropical Forest Science and Forest Global Earth Observatories (CTFS-ForestGEO)' co-ordinated by the Smithsonian Tropical Research Institute, which uses a standard protocol for large forest plots (2–60 hectares). These now form a unified, global network for the study of tropical and temperate forest function and diversity. The multi-institutional network comprises over 60 forest research plots across the Americas, Africa, Asia and Europe, with a strong focus on tropical regions. It monitors the growth and survival of approximately 6 million trees and 10,000 species that occur in the forest plots.[1] Although not primarily targeted at plant conservation, the tropical plots in particular have made important contributions to plant species inventory and monitoring. Several botanic gardens have been involved in this important research programme, including XTBG and South China Botanic garden in China.

Setting up large, CTFS-standard forest plots is beyond the capacity of many botanic gardens, but smaller plots can also be useful for conservation monitoring, particularly for herbs and shrubs, and are an important supplement to the large plots which, necessarily, are sparsely distributed and thus exclude numerous rare species. Even then, all plots are samples, and as complete inventories of regions are not a realistic option for the foreseeable future, the design of data-collecting activities should be based soundly in ecological theory and should enable the application of proven statistical techniques to the modelling of wider spatial distribution patterns from the point locations that these field records were taken from (Margules and Pressey, 2000).

Almost all existing monitoring is demographic, but small, isolated, populations of threatened species are also vulnerable to loss of genetic diversity. Moreover, widespread species often show genetic structure which, at least in part, reflects local adaptation, so conservation may need to target multiple populations spread across the distribution range. A variety of molecular markers have been used for genetic monitoring in the past, but in the last few years the use of next generation sequencing has increased resolution and accuracy, while decreasing costs (Corlett, 2016c). Few botanic gardens can do this 'in house', but outsourcing to commercial providers is common in conservation-related studies and can include much of the data analysis.

A variety of new technologies can help with monitoring, particularly in large or remote areas where regular field surveys may be impractical (Marvin *et al.*, 2016). Remote sensing from satellites is widely used for monitoring vegetation change. Data from a variety of satellite sensors is freely available, while very high spatial resolutions are available from commercial satellites if funding is available. Assessing this data is becoming easier via new, web-based platforms, and Global Forest Watch now provides real-time monitoring of forests across the tropics (see Box 4.5). The use of autonomous drones (unmanned aircraft systems, UAS)

Box 4.5 *Real-Time Monitoring of Forest Loss in Tropics*

Mongabay is a conservation news website with a special focus on forests, including issues like conservation, deforestation and forest degradation. For much of its history, Mongabay relied on data from the UN Food and Agriculture Organization (FAO) to track trends in forest cover across regions and countries. Subnational forest monitoring was not possible using FAO data, which was often inaccurate since it is self-reported by governments.

In 2011, Mongabay partnered with the Carnegie–Ames–Stanford Approach (CASA) ecosystem modelling team at NASA Ames Research Center to develop the Global Forest Disturbance Alert System (GloFDAS). The tool provides data on forest disturbance globally on a quarterly basis using NASA's QUICC product, which is based on comparison of MODIS global vegetation index images at the exact same time period each year in consecutive years. GloFDAS shows potential forest disturbance locations as centre points of 5-kilometre by 5-kilometre areas that were detected with a significant loss of forest greenness cover (greater than 40 per cent) over the previous 12 months, alerting users to potential deforestation. GloFDAS gave Mongabay the unprecedented ability to independently monitor forest change on a global basis, allowing journalists to identify potential deforestation within three months of occurrence. As a result, Mongabay used the tool extensively in fact-checking claims around deforestation and to identify new deforestation hotspots. The utility of GloFDAS was recognised in 2013 when it was honoured as a finalist for the Katerva Award.

In February 2014, the World Resources Institute launched the second iteration of Global Forest Watch, a forest monitoring platform that offers capabilities that extend well beyond GloFDAS. Global Forest Watch incorporated a wide range of data on forests into its interactive maps. Represented as 'layers', this data includes forest gain and loss, concession data, tree height and carbon density, protected areas, fire data and more. The platform allows users to download all data as shape files or tabular spreadsheets. Critically, Global Forest Watch extended bi-weekly forest monitoring across the humid tropics with its FORMA alert system, which leverages NASA MODIS satellite imagery at a 250 m by 250 m resolution. The FORMA system enabled anyone with an Internet connection to see what was happening in the world's tropical forests on a near-real-time basis, providing a powerful tool in efforts to arrest deforestation.

For an environmental journalism organisation like Mongabay, Global Forest Watch serves as a revolutionary tool for fact-checking claims on forests. Journalists can quickly evaluate whether NGOs, governments and companies are accurately representing what is occurring in forests. Journalists can also use the alert functionality to detect when forest is being cleared within a given area, providing a new source of potential story ideas. In 2016, Global Forest Watch launched GLAD alerts, which track forest cover at a 30 m by 30 m resolution, enabling users to detect forest degradation on a near-real-time basis for the first time.

Source: Rhett Butler, the founder of Mongabay

increasingly fills the gap between remote sensing by satellite and ground surveys (Marvin *et al.*, 2016). GPS telemetry can be used to track animal movements remotely while increasingly sophisticated digital camera traps provide a range of data on animal presence, movement and behaviour, as well as illegal human activities. Metabarcoding can provide quicker, cheaper and more accurate assessment of biodiversity in groups, such as most invertebrates, that are difficult to assess by traditional methods (Corlett, 2016c; see Chapter 2). A major challenge for the future will be integrating these technologies with traditional field observations to provide real-time support for protected area management (Marvin *et al.*, 2016).

4.7 Ecological Restoration of Degraded Areas

In human-dominated landscapes, protection of existing vegetation is rarely enough: active ecological restoration is needed to restore, buffer and link these areas in order to maximise their conservation value. The unidirectional environmental changes that characterise the Anthropocene have put the traditional goal of recreating replicas of past ecological communities out of reach, but the creation of new communities resilient to current and future conditions is still an important goal (Corlett, 2016a). Many botanic gardens are involved in ecological restoration, utilising their full range of conservation, horticultural and social skills. Understanding how the composition of the ecological community affects the long-term sustainability and ecosystem functioning of restoration efforts is also important. Knowledge about the successional status of different species and their ability to recreate mature vegetation and functioning communities is powerful. The framework species method for forest restoration (Elliott *et al.*, 2014) utilises local knowledge of species regeneration to establish a core population of trees that create a closed canopy quickly, which then allows natural processes of seed dispersal to further diversify the forests. Additionally, long-term experiments have been established at the Morton Arboretum to understand the impact of the deep evolutionary diversity on long-term dynamics of prairie communities. Is species richness a sufficient criteria or does the relatedness of those species and how much of the tree of life of plants is included in the restoration effort matter? These experiments are using carefully designed plantings of native plants to assess the impact of the phylogenetic diversity of plants on the eventual outcome of restoration on soils, ecosystem functions and community composition and diversity (Hipp *et al.*, 2015).

4.8 Assisted Migration in Response to Climatic Change

The ongoing anthropogenic climate change has been predicted to result in an accelerating rate of extinctions. A recent review of published studies estimated that 7.9 per cent of plant species will become extinct as a result of climate change (Urban, 2015). Plants' responses to climate change may include migration, *in situ* acclimation and adaptation and extirpation (Christmas *et al.*, 2016). Migration to track climate change in space is likely to be limited by the dispersal abilities of many plant species (Corlett and Westcott, 2013), so human-assisted migration is one possible conservation response. Moving species outside their native ranges in anticipation of future climatic conditions is – rightly – controversial, and should only be done after all potential costs and benefits have been investigated. For plants, the combination of ecological and horticultural skills with *ex situ* collections is likely to make botanic gardens among those most capable organisations for implementing this practice (Heywood, 2011). A variety of modelling approaches have been used to assess the need for assisted migration (Koralewski *et al.*, 2015; Hăllfors *et al.*, 2016), but simple models based only on abiotic factors may sometimes give misleading results. For example, a study of Fremont cottonwood (*Populus fremontii*) found that species resistance to cold may interact with pathogen resistance, and the authors suggest applying multiple and intermediate transfer phases to supplement local genetic variation for effective assisted migration (Grady *et al.*, 2015). Specific recommendations for botanic gardens in plant assisted migration research and practice are listed in Box 4.6.

4.9 Ethnobotany and the Sustainable Use of Natural Resources

The conventional 'fences and fines' approach to *in situ* conservation can lead to conflicts with local communities who lose access to resources that they have used for generations. Villages may be moved, grazing restricted, and hunting and plant collection banned. Animals from protected areas may damage crops and kill or injure people who live nearby, as has happened with Asian elephants in Southwest China. Attempts at compromise have generally involved zoning, so that local people have

> **Box 4.6** *Recommendation for Botanic Gardens in Plant Assisted Migration Research and Practice*
>
> - Assess the climate-change vulnerability of the regional flora using appropriate modelling techniques.
> - Use a common garden approach to understand the responses of plant material from different populations to climate change (Aitken and Bemmels, 2015).
> - Assess the opportunities for enhancing the adaptability of threatened populations by reinforcement with climate-adapted genotypes from other populations or *ex situ* collections.
> - Conduct carefully monitored experiments with managed translocations.
> - Enhance the connectivity of the protected area system, by expanding the area under protection and protecting or restoring corridors, in order to maximise the opportunities for spontaneous plant movements in response to climate change.

access to certain resources in certain parts of the protected area, designated for sustainable use. Such partial protection is more difficult to manage than complete exclusion, but there are enough successful examples to suggest that this approach can work. Research on ethnobotany – a traditional focus of botanic gardens – can potentially help in the design of sustainable management practices. A list of recommended actions for botanic gardens in promoting sustainable use of natural resources by local people in protected areas for poverty alleviation is in Box 4.7.

4.10 Set up Partnerships with All Stakeholders

In situ conservation is a comprehensive programme which often involves many different participants. Governmental agencies are usually responsible for the design, establishment and management of protected areas. Local communities are another stakeholder in conservation. In many parts of the world, local people rely on hunting and gathering forest products for making their living, and setting up protected areas may threaten these activities. On the other hand, many studies have shown that, without support from the local community, it is very difficult to achieve conservation goals. Other participants in *in situ* conservation may

> Box 4.7 *Ways in which Botanic Gardens can Help Local Communities to Benefit from* In Situ *Conservation*
>
> - Provide technical support and training to local people for harvesting NTFPs sustainably from protected areas.
> - Assist with marketing of sustainably harvested wild products.
> - Support local people and their rights in negotiations with protected area authorities.
> - Help with promotion of ecotourism and ensuring that local people's interests are protected.
> - Conduct outreach for explaining the multi-values including ecosystem service, health and social well-being brought by protected areas to local people.

include NGOs, universities and other institutions with conservation-related projects in the area.

The majority of people now live in cities and this trend is continuing to increase. Most botanic gardens are located in or near cities, and many are engaged with local urban communities in the management of urban vegetation for both recreation and conservation. In the context of *in situ* conservation, botanic gardens can promote a diversity of trees and other plants, favouring species native to the region, and utilising their horticultural skills and experience to develop ways of using endangered species in urban plantings. The Chicago Region Trees Initiative (CRTI), founded by the Morton Arboretum, is a collaboration of leading organisations and partners from the seven-county Chicago metropolitan area, working to preserve, protect and enhance the health of the urban forest in order to boost the quality of life for humans and wildlife. The CRTI has assembled a network of over 250 organisations including landowners, land managers, industry and nurseries, and works at a grassroots level, to educate and promote good urban forestry practices.[2] The CRTI has compiled an extensive GIS database and is collaborating with scientists at the Field Museum and the University of Chicago to create an accurate, high resolution, and useful map of the urban forest of the Chicago region to help guide policy on where trees will have the most positive impact and what types of trees should be planted. Cities around the world are making efforts to improve the living condition of their human populations and botanic gardens should explore ways in which to enhance this process.

The need for botanic gardens' inputs into *in situ* biodiversity conservation is huge, while the capacities of individual botanic gardens are limited (Kramer *et al.*, 2013). Particularly in biodiversity-rich areas, therefore, botanic gardens need to develop partnerships with all the sectors. The permanence and stability of botanic gardens, their unique resources, and their typically excellent relationships with local governments and communities, make them natural aggregators of inputs and, where appropriate, leaders. Such partnerships may generate outputs that are much beyond expectations. There are numerous examples of this, including XTBG's own 10-year cooperation with the local nature reserve administrative bureau, which has resulted in significant outcomes in promoting biodiversity research and conservation in the biodiversity hotspot (see Box 4.8). Another example, highlighting south–north cooperation, is that of Missouri Botanic Garden working with local communities in Madagascar (see Box 4.9).

Box 4.8 *Ten Years' Cooperation between XTBG and the Xishuangbanna National Natural Reserve Administrative Bureau (XNNRAB)*

Xishuangbanna is a biodiversity hotspot, harbouring more than one-seventh of China's native plant species in its 19,000 km^2, which is only 1/500 of China's territory in area. Due to the rapid population increase and especially the expansion of rubber plantations (Qiu, 2009), biodiversity loss at an alarming rate has been recognised as one of the serious environmental issues in this biodiversity hotspot.

The Xishuangbanna Tropical Botanic Garden (XTBG) is a national institute for biodiversity conservation that is affiliated to the Chinese Academy of Sciences. Since its establishment in 1959, XTBG has taken conserving biodiversity in Xishuangbanna as its research priority. The Xishuangbanna National Natural Reserve Administrative Bureau (XNNRAB) is the administrative body in charge of the national natural reserve in Xishuangbanna, which comprises five separately distributed pieces of forest, totalling 2,400 km^2, 12 per cent of total area in Xishuangbanna. The XTBG and XNNRAB are the two main agencies devoted to conservation. From 2006, the two organisations started to set up a strategic partnership and decided to work together to create collective efforts for conservation of the area.[3] An annual meeting is held with the participation of conservation scientists and senior staff to agree on cooperative projects, who will be in charge of them and a final budget agreed by both sides. After 10 years' cooperation, it has delivered a large number of significant achievements in conservation in the region. For example:

- Jointly setting up two new protected areas (named Bulong protected area and Yiwu protected area) increasing the total protected area in Xishuangbanna from 14 per cent to 18 per cent.
- Jointly conducting a systematic survey on orchids, providing the most up to date and comprehensive information for the conservation status of orchids in Xishuangbanna (Liu et al., 2015).
- Jointly setting up long-term monitoring plots including one CTFS standard 25 ha plot, one canopy crane and 20 1-ha plots across the whole region.
- Conducting a wildlife survey using camera traps and monitoring across the whole protected area.
- Providing forecasts (with the assistance of XTBG), of the occurrence of the wild elephant population and significantly reducing injuries to local people by elephants.
- Jointly conducting large-scale outreach for conservation education to all the schools surrounding the protected areas.

Box 4.9 *Missouri Botanic Garden Working with Local Communities in Madagascar*

Madagascar is renowned for its rich flora with over 80 per cent endemism. Over the past 40 years Missouri Botanic Garden has helped to comprehensively document the Malagasy flora. The Madagascar Catalogue project is reviewing the taxonomic framework for each plant genus and is determining the species names together with their synonyms. Associated data on geographic distribution, ecology and conservation status is recorded for each accepted species. So far IUCN Red List categories and criteria have been applied to about one-third of the endemic plant species.

The plant conservation data collected by Missouri botanists is essential for conservation planning on the island. Based on GIS data on plant diversity and analysis of the distribution of more than 1,200 endemic species, a map of 77 sites critical for plant conservation in Madagascar was produced. Missouri Botanic Garden worked closely with Royal Botanic Gardens, Kew on this important project. The government of Madagascar granted temporary protection status for 25 of these sites within the national protected area system.

The highly restricted littoral forest of eastern Madagascar has been a particular target for conservation action. The species-rich rainforest adapted to the sandy soils runs parallel to the east coast of the island but only around 20 per cent of the forest remains. About 1,200 plant species, or 10 per cent of Madagascar's flora, have been recorded from the remnant forests and about half these plants are found only in this habitat type. Missouri Botanic Garden has identified the 15 most important sites for protection within this narrow strip of coastal forest, and these are included within the 77 national priority areas. Now the Garden staff are working to help communities in two areas of littoral forest, to manage the forest resources sustainably.

Missouri Botanic Garden is currently managing 13 conservation sites in Madagascar, 12 of which are formally protected areas, and has initiated forest regeneration projects with 11 local communities on sites encompassing 228 square miles. The restoration sites represent a wide range of native vegetation that will be designated as protected areas. Efforts are underway with local communities to improve livelihoods, develop alternative sources of food and fuelwood, and improve income generation, so that pressures on the forest are reduced.

4.11 Expanding Impact by Networking

Scaling up conservation efforts is often difficult. A single botanic garden's capabilities are necessarily limited, especially from the perspective of *in situ* conservation, which has rarely been considered a core responsibility. The botanic garden community, however, has been very successful in extending impact by networking, both regionally and globally. Indeed, a recent study showed that botanic gardens which are members of a global botanic garden network are often implementing more targets of the GSPC (Williams *et al.*, 2012).

One example from the global botanic garden community is the Ecological Restoration Alliance of Botanic Gardens (ERA).[4] The ERA is a global consortium of botanic gardens actively engaged in ecological restoration. As highlighted in Chapter 6, members of the alliance have agreed to support efforts to scale up the restoration of damaged, degraded and destroyed ecosystems around the world, supporting the UN target to restore 15 per cent of the world's degraded ecosystems by 2020. Another, regional, example is the recently created Chinese Union of Botanic Gardens (CUBG). The CUBG is a new initiative formally announced by the Chinese Academy of Sciences, with support from the State Forestry Administration of China and the Ministry of Housing and Urban–Rural Construction. It aims to coordinate the collective efforts of botanic gardens for conservation throughout China. Currently there are 94 member institutions in CUBG.[5] The Zero Extinction Project, a programme developed by XTBG to prevent local plant extinctions by using all available conservation tools, has now been extended to eight different regions throughout China, covering in total 28 per cent of China's territory.[6]

4.12 Challenges and Opportunities

As highlighted in the introduction to the book and this chapter, many environmental scientists believe the planet has entered into a new geological epoch – the Anthropocene – in which human influences are at least as important as the natural environment (Steffen *et al.*, 2011). 'Business as usual' is no longer possible for conservation (Corlett, 2015). Although it is still true that 'there is no technical reason why any plant species should become extinct',[7] achieving this in practice for tens of thousands of threatened plant species is increasingly challenging under this new scenario. Protecting areas and restoring degraded vegetation is no longer enough when even undisturbed, 'pristine' areas are subject to the impacts of ongoing global change: changing climate, rising carbon dioxide levels, regional air pollution, increasing nutrient deposition, biological invasions and others (Krupnick, 2013; Corlett, 2015). Additionally, because of global trade and the greatly increased flow of natural goods around the world, no place remains 'remote'. Future waves of invasive pests, diseases and species should be expected to impact local, native communities around the world. Given these current trends and realities, *in situ* conservation activities need to embrace all vegetation, from the most natural to the entirely artificial. In the Anthropocene, plants must acclimate, adapt, move or die, and conservationists must predict which species will require human assistance to survive, and then identify the most appropriate form of management intervention. Interestingly, plants appear to experience the 'great extinctions' differently than animals (Silvestro et al., 2015), indicating that something fundamental about plant biology allows lineages to avoid extinction. Botanists should attempt to understand and exploit this aspect of plant evolutionary biology.

This challenge calls for both research and action. Botanic garden capacity and commitment to conservation varies considerably, but a recent survey of 255 of them from 67 countries showed that botanic gardens from both the 'old north' and the 'new south' are committed to the implementation of the GSPC (Williams et al., 2012). The Anthropocene calls for a more integrative approach, in which biodiversity conservation, ecosystem sustainability and human wellbeing are included under the same umbrella (see Box 4.10). *In situ* conservation will undoubtedly form a core part of these efforts.

Box 4.10 *Recommendations for Botanic Gardens Involving* In Situ *Conservation*

- Strengthen skills and knowledge on restoration ecology and develop local habitat recovery programmes.
- Investigate the mechanisms and conditions that facilitate and promote the invasiveness of exotic plant species.
- Preserve indigenous knowledge of ecosystem management and incorporate into conservation plans.
- Study the market-oriented mechanisms for biomass and biodiversity conservation and take advantage of emerging opportunities.
- Seek the latest and most effective technologies, including bioinformatics, GIS, Internet resources and genomics, and develop the necessary expertise to effectively utilise them.
- Promote 'south–north' cooperation within the botanic garden community to maximise conservation efforts.
- Develop close reciprocal ties with government officials, from local to national levels.
- Cooperate and collaborate directly with large-scale industries, from energy to agriculture to tourism, that affect natural resource use.
- Provide leadership in climate change research and education, and explore the feasibility of managed relocation of species most vulnerable to extinction owing to climate change, providing successful

cases for saving the critically endangered species via assisted migration.
- Explore the mechanism and develop a model for integrating all conservation tools in the conservation toolbox into zero plant extinction at both local and regional scales.
- Work with local people and all the stakeholders for both conservation and sustainable development and social wellbeing in the target area.

Source: after Chen *et al.*, 2009 with modification

4.13 Conclusions

The international community has set ambitious targets for plant conservation (CBD, 2010), but these are unlikely to be achieved by the 2020 deadline, despite increasing efforts (Tittensor *et al.*, 2014). A recent global estimate, based on a random sample of 7,000 species, is that 22 per cent of land plant species are threatened (Brummitt *et al.*, 2015). A model-based assessment of 15,200 Amazonian tree species estimated that 36–57 per cent would qualify as threatened under red list criteria, suggesting that tropical trees may be among the most threatened organisms on Earth (ter Steege *et al.*, 2015). Percentages like these underline the fact that no single approach, either *in situ* or *ex situ*, can deal with the problem of conserving global plant diversity. This will require the persistent, integrated use of the entire conservation toolbox.

Botanic gardens are well placed to play a major role in this, as shown in this chapter (and others in this book). With regard to *in situ* conservation, they can provide technical and resource support, making use of their expertise in taxonomy, ecology, horticulture and other fields, as well as their long-term records and *ex situ* germplasm collections. Botanic gardens can also act as a platform for bringing together efforts in conservation and sustainable development (Blackmore *et al.*, 2011). In the tropics, in particular, they can and should take a leading role in plant conservation efforts, while major botanic gardens outside the tropics need to continue their long tradition of 'south–north' cooperation, so that they act as a conduit for new ideas and techniques into local conservation activities. In conclusion, every botanic garden needs to look at its location, its collections and its expertise, and ask: how can we contribute most effectively to the conservation of global plant diversity?

Notes

1. See www.forestgeo.si.edu/.
2. See www.chicagorti.org.
3. See www.xtbg.cas.cn/xwzx/zhxw/201512/t20151217_4498351.html.
4. See www.erabg.org/index/.
5. See www.cubg.cn.
6. See www.cubg.cn/work/conservation/.
7. See www.bgci.org/about-us/5yrplan/.

References

Aitken, S. N. and Bemmels, J. B. (2015). Time to get moving: assisted gene flow of forest trees. *Evolutionary Application*, 9: 271–290.

Ali, N. S. and Trivedi, C. (2011). Botanic gardens and climate change: a review of scientific activities at the Royal Botanic Gardens, Kew. *Biodiversity and Conservation*, 20: 295–307.

Baldock, K. C. R., Goddard, M. A., Hicks, D. M. et al. (2015). Where is the UK's pollinator biodiversity? The importance of urban areas for flower-visiting insects. *Proceedings of the Royal Society B: Biological Sciences*, 282 (1803): 20142849.

Bernstein, M. J., Wiek, A., Brundiers, K., Pearson, K., Minowitz, A., Kay, B. and Golub, A. (2016). Mitigating urban sprawl effects: A collaborative tree and shade intervention in Phoenix, Arizona, USA. *Local Environment*, 21: 414–431.

Blackmore, S., Gibby, M. and Rae, D. (2011). Strengthening the scientific contribution of botanic gardens to the second phase of the global strategy for plant conservation. *Botanical Journal of the Linnaean Society*, 166: 267–281.

Brummitt, N. A., Bachman, S. P., Griffiths-Lee, J. et al. (2015). Green plants in the red: a baseline global assessment for the IUCN Sampled Red List Index for Plants. *PloS ONE*, 10(8), e0135152.

Butchart, S. H. M., Walpole, M., Collen, B. et al. (2010). Global biodiversity: indicators of recent declines. *Science*, 328(5982): 1164–1168.

Chen, J., Cannon, C. H. and Hu, H. B. (2009). Tropical botanic gardens: at the *in situ* ecosystem management frontier. *Trends in Plant Science*, 14: 584–589.

Christmas, M. J., Breed, M. F. and Lowe, A. J. (2016). Constraints to and conservation implications for climate change adaptation in plants. *Conservation Genetics*, 17: 305–320.

Convention on Biological Diversity (CBD). (2010). Consolidated update of the global strategy for plant conservation 2011–2020. Available online at www.cbd.int/gspc/strategy.shtml [accessed March 2017].

Corlett, R. T. (2015). The anthropocene concept in ecology and conservation. *Trends in Ecology and Evolution*, 30: 36–41.

Corlett, R. T. (2016a). Restoration, reintroduction, and rewilding in a changing world. *Trends in Ecology and Evolution*, 31: 453–462.

Corlett, R. T. (2016b). Plant diversity in a changing world: status, trends, and conservation needs. *Plant Diversity*, 1: 11–18.

Corlett, R. T. (2016c). A bigger toolbox: biotechnology in biodiversity conservation. *Trends in Biotechnology*, 35(1): 55–65.

Corlett, R. T. and Westcott, D. A. (2013) Will plant movements keep up with climate change? *Trends in Ecology and Evolution*, 28: 482–488.

Dirzo, R., Young, H. S., Galetti, M., Ceballos, G., Isaac, N. J. B. and Collen, B. (2014). Defaunation in the Anthropocene. *Science*, 345: 401–406.

Elliott, S., Blakesley, D. and Hardwick, K. (2014). *Restoring Tropical Forests: A Practical Guide*. Royal Botanic Gardens, Kew, UK: Kew Publishing.

Folke, C., Holling, C. S. and Perrings, C. (1996). Biological diversity, ecosystems, and the human scale. *Ecological Applications*, 6: 1018–1024.

Fotinos, T. D., Namoff, S., Lewis, C., Maschinski, J., Griffith, M. P. and von Bettberg, E. J. B. (2015). Genetic evaluation of a reintroduction of Sargent's cherry palm. *Journal of the Torrey Botanical Society*, 142: 51–62.

Grady, K. C., Kolb, T. E., Ikeda, D. H. and Whitham, T. G. (2015). A bridge too far: cold and pathogen constraints to assisted migration of riparian forests. *Restoration Ecology*, 23: 811–820.

Haddad, N. M, Brudvig, L. A., Clobert, J. et al. (2015). Habitat fragmentation and its lasting impact on Earth's ecosystems. *Science Advances*, 1: e1500052.

Hällfors, M. H., Aikio, S., Fronzek, S., Hellmann, J. J., Ryttäri, T. and Heikkinen, R. K. (2016). Assessing the need and potential of assisted migration using species distribution models. *Biological Conservation*, 196: 60–68.

Harrison, R. D., Tan, S., Plotkin, J. B., Slik, J. W. F., Detto, M., Brenes, T., Itoh, A. and Davies, S. J. (2013). Consequences of defaunation for a tropical tree community. *Ecology Letters*, 16: 687–694.

Heywood, V. H. (2011). The role of botanic gardens as resource and introduction centres in the face of global change. *Biodiversity and Conservation* 20: 221–239.

Hipp, A. L., Larkin, D. J., Barak, R. S., Bowles, M. L., Cadotte, M. W., Jacobi, S. K., Lonsdorf, E., Scharenbroch, B. C., Williams, E. and Weiher, E. (2015). Phylogeny in the service of ecological restoration. *American Journal of Botany*, 102: 647–648.

Holtgrieve, G. W., Schindler, D. E., Hobbs, W. O. et al. (2011). A coherent signature of anthropogenic nitrogen deposition to remote watersheds of the northern hemisphere. *Science*, 334: 1545–1548.

Koralewski, T. E., Wang, H.-H., Grant, W. E. and Byram, T. D. (2015). Plants on the move: assisted migration of forest trees in the face of climate change. *Forest Ecology and Management*, 344: 30–37.

Kramer, A. T., Barbara, Z.-A. and Havens, K. (2013). Botanical capacity assessment project to achieve 2020 global strategy for plant conservation targets. *Annals of the Missouri Botanic Garden*, 99: 172–179.

Krupnick, G. A. (2013). Conservation of tropical plant biodiversity: what have we done, where are we going? *Biotropica*, 45: 693–708.

Laurance, W. F. et al. (2012). Averting biodiversity collapse in tropical forest protected area. *Nature*, 489: 290–294.

Le Saout, S., Hoffmann, M., Shi, Y. et al. (2013). Protected areas and effective biodiversity conservation. *Science*, 342: 803–805.

Leverington, F., Costa, K. L, Pavese, H., Lisle, A. and Hockings, M. (2010). A global analysis of protected area management effectiveness. *Environmental Management*, 46(5): 685–698.

Liu, Q., Chen, J., Corlett, R. T., Fan, X. L., Yu, D. L., Yang, H. P. and Gao, J. Y. (2015). Orchid conservation in the biodiversity hotspot of southwestern China. *Conservation Biology*, 29(6): 1563–1572, doi: 10.1111/cobi.12584.

MacGregor-Fors, I., Escobar, F., Rueda-Hernández, R. et al. (2016). City 'green' contributions: the role of urban greenspaces as reservoirs for biodiversity. *Forests, Trees and Livelihoods*, 7: 146.

Mammides, C., Goodale, U. M., Corlett, R. T. et al. (2016). Increase geographic diversity in the international conservation literature: a stalled process? *Biological Conservation*, 198: 78–83.

Margules, C. R. and Pressey, R. L. (2000). Systematic conservation planning. *Science*, 405: 243–253.

Marignani, M. and Blasi, C. (2012). Looking for important plant areas: selection based on criteria, complementarity, or both? *Biodiversity and Conservation*, 21: 1853–1864.

Marvin, D. C., Koh, L. P., Lynam, A. J., Wich, S., Davies, A. B., Krishnamurthy, R., Stokes, E., Starkey, R. and Asner, G. P. (2016). Integrating technologies for scalable ecology and conservation. *Global Ecology and Conservation*, 7: 262–275.

Pimm, S. L. and Joppa, L. N. (2015). How many plant species are there, where are they, and at what rate are they going extinct? *Annals of the Missouri Botanic Garden*, 100: 170–176.

Pinto, M., Almeida, C., Pereira, A. M. and Silva, M. (2016). Urban Forest Governance: FUTURE – The 100,000 Trees Project in the Porto Metropolitan Area. In: Castro, P., Azeiteiro, U. M., Bacelar-Nicolau, P., Filho, W. L. and Azul, A. M. (Eds), *Biodiversity and Education for Sustainable Development*. Switzerland: Springer International Publishing, pp. 187–202.

Primack, R. B. and Miller-Rushing, A. J. (2009). The role of botanic gardens in climate change research. *New Phytologist*, 182: 303–313.

Qiu, J. (2009). Where the rubber meets the garden. *Nature*, 457: 246–247.

RBG Kew (2016). *The State of the World's Plants Report 2016*. Royal Botanic Gardens, Kew, available online at https://stateoftheworldsplants.com/areas-important-for-plants [accessed February 2017].

Ren, G. P., Young, S. S., Wang, L., Wang, W., Long, Y. C., Wu, R. D., Li, J. S., Zhu, J. G. and Yu, D. W. (2015). Effectiveness of China's National Forest Protection Program and nature reserves. *Conservation Biology*, 29: 1368–1377.

SCBD (2010). *COP 10 Decision X/2: Strategic Plan for Biodiversity 2011–2020.*, Nagoya, Japan: Secretariat of the Convention on Biological Diversity.

Secretariat of the Convention on Biological Diversity (CBD) (2008). *Protected Areas in Today's world: Their Values and Benefits for the Welfare of the Planet*. Montreal, CBD Technical Series No. 36, i–vii + 96 pp.

Sharrock, S., Oldfield, S. and Wilson, O. (2014). *Plant Conservation Report 2014: A Review of Progress in Implementation of the Global Strategy for Plant Conservation 2011–2020*. Richmond, UK: Secretariat of the Convention on Biological Diversity and Botanic Gardens Conservation International.

Silvestro, D., Cascales-Miñana, B., Bacon, C. D. and Antonelli, A. (2015). Revisiting the origin and diversification of vascular plants through a comprehensive Bayesian analysis of the fossil record. *The New Phytologist*, 207(2): 425–436.

Soulsbury, C. D. and White, P. C. L. (2016). Human–wildlife interactions in urban areas: a review of conflicts, benefits and opportunities. *Wildlife Research*, 42: 541–553.

Steffen, W., Grinevald, J., Crutzen, P. and McNeill, J. (2011). The Anthropocene: conceptual and historical perspectives. *Philosophical Transactions. Series A, Mathematical, Physical, and Engineering Sciences*, 369(1938): 842–867.

ter Steege, H., Pitman, N. C. A., Killeen, T. J. *et al.* (2015). Estimating the global conservation status of more than 15,000 Amazonian tree species. *Science Advances*, 1(10): e1500936.

Tittensor, D. P. *et al.* (2014). A mid-term analysis of progress toward international biodiversity targets. *Science*, 346: 241–244.

Urban, M. C. (2015). Accelerating extinction risk from climate change. *Science*, 348: 571–573.

Van der Burgt, X. M. *et al.* (2015). The *Gilbertiodendron ogoouense* species complex (Leguminosae: Caesalpinioideae), Central Africa. *Kew Bulletin*, 70: 29.

Vovides, A. P., Iglesias, C., Luna, V. and Balcazar, T. (2013). Botanic gardens and the biodiversity crisis. *Botanical Sciences*, 91: 239–250.

Wang, C. J., Wan, J. Z., Zhang, G. M., Zhang, Z. X. and Zhang, J. (2016). Protected areas may not effectively support conservation of endangered forest plants under climate change. *Environmental Earth Sciences*, 75: 466, doi: 10.1007/s12665-016-5364-4.

Williams, M., Zalasiewicz, J.,Waters, C. N. *et al.* (2016). The Anthropocene: a conspicuous stratigraphical signal of anthropogenic changes in production and consumption across the biosphere. *Earth's Future*, 4: 34–53.

Williams, S. J., Jones, J. P. G., Clubbe, C., Sharrock, S. and Gibbons, J. M. (2012). Why are some biodiversity policies implemented and others ignored? Lessons from the uptake of the global strategy for plant conservation by botanic gardens. *Biodiversity and Conservation*, 21: 175–187.

Wyse Jackson, P. S. and Kennedy, K. (2009). The global strategy for plant conservation: a challenge and opportunity for the international community. *Trends in Plant Science*, 14: 578–580.

Zhang, M. G., Zhou, Z. K., Chen, W. Y., Cannon, C. H., Raes, N. and Slik, J. W. F. (2014). Major declines of woody plant species ranges under climate change in Yunnan, China. *Diversity and Distributions*, 20: 405–415.

Zhang, Y. B., Wang, Y. Z., Zhang, M. J. and Ma, K. P. (2014). Climate change threats to protected plants of China: an evaluation based on species distribution modelling. *Chinese Sciences Bulletin*, 59: 4652–4659.

5 · *The Role of Botanic Gardens in* Ex Situ *Conservation*

PAUL SMITH AND VALERIE PENCE

5.1 Introduction: The Range of Plant Diversity Held by Botanic Gardens in their Living Collections

In this chapter we look at the value of botanic garden collections, data, infrastructures and expertise in addressing the targets of the Global Strategy for Plant Conservation and the conservation and sustainable use of biodiversity more broadly. Botanic garden collections comprise living collections in botanic garden landscapes, nurseries, glasshouses and micro-propagation laboratories; seed collections in seed banks; cryopreserved collections; pressed plant specimens in herbaria; DNA and RNA collections in molecular banks; fungal collections; pollen collections; wood sample collections; spore banks; artefact collections; art collections; libraries and archives and so on. All of these collections and their associated data represent a tremendous resource for plant conservation.

As a professional community, botanic gardens and arboreta conserve and manage a far greater proportion of known plant species diversity than forestry, agriculture or any other sector. BGCI's PlantSearch database includes 1.3 million accession names from 1,141 botanic gardens around the world.[1] A recent comparison with *The Plant List* (2013) indicates that those gardens manage at least 115,787 different species in their living collections – equivalent to 33 per cent of all the species listed in *The Plant List* (Smith, 2016). There are, of course, caveats such as the fact that accession records are not always up to date or accurately named. However, this can be balanced against the fact that PlantSearch itself is not comprehensive, covering only about 40 per cent of all botanic gardens. The detailed records are possible because of the infrastructures, expertise and methods employed by botanic gardens for managing plant diversity.

A key defining criterion for botanic garden status, as opposed to a public park, for example, is *well-documented collections*. Botanic garden collections are named, labelled and accompanied by comprehensive data, including their origin, date of collection, collection locality, and so on. This documentation of collections is crucial in order for the botanic gardens community collectively to add to the knowledge base for any particular species – how to propagate it, its pathology, abiotic tolerances and other factors.

The botanic garden sector includes world-leading infrastructures. Kew's Millennium Seed Bank, the Royal Botanic Garden, Sydney's Plant Bank and Kunming Institute of Botany's Gene Bank of Wild Species are the largest, most sophisticated seed banks in the world. The sector is equally strong in glasshouse and horticulture infrastructures and well served with micro-propagation facilities and molecular laboratories. The botanic garden community's most comprehensive data source on garden facilities and research foci is BGCI's GardenSearch,[2] a web-based register of the world's botanic gardens comprising information on 2,671 botanic gardens and arboreta in 135 countries.

At the heart of the botanic gardens network, of course, are the skills and knowledge of staff. BGCI's GardenSearch database indicates that the world's botanic gardens employ at least 60,000 people, comprising thousands of plant scientists and horticulturalists who possess unique knowledge right across the taxonomic spectrum. This expertise is also manifest in specialist networks covering plant conservation practice such as red listing, seed conservation and ecological restoration.

In fact, the botanic garden community has been so successful in conserving and growing plant diversity, *ex situ*, that it has postulated that there is no technical reason why any plant species should become extinct (Smith *et al.*, 2011). Given the array of *ex situ* and *in situ* conservation techniques employed by the botanic garden community (including species conservation status assessment, seed banking, cultivation, tissue culture, cryopreservation, assisted migration, species recovery and ecological restoration) we should be able to avoid species extinctions.

Botanic gardens have an important role to play in achieving most, if not all, of the GSPC targets but perhaps the key target for botanic gardens is Target 8: *At least 75 per cent of threatened plant species in* ex situ *collections, preferably in the country of origin, and at least 20 per cent available for recovery and restoration programmes.*

Estimates of the proportion of threatened plant species in *ex situ* collections in botanic gardens vary with the definition of 'threatened', the location and the availability of information. In Australia and New

Zealand, it is estimated that 56 per cent of threatened plant diversity is held in *ex situ* living collections or seed banks (BGCI, 2013a). In Europe, based on comparing PlantSearch records with Europe's Red List of Threatened Plants, the figure is 71 per cent with 50.5 per cent held in seed banks (ENSCONET, 2015); however, not all European plant species have been assessed for their conservation status. The Russian Federation of Botanic Gardens, which includes gardens in former Soviet republics, estimates that 64 per cent of threatened plant diversity is conserved *ex situ* (Demidov, 2012). Finally, in North America it is estimated that 39 per cent of threatened plant taxa are held *ex situ* (BGCI, 2013b).

While these figures tell us little about the quality of the collections of threatened species or for that matter their conservation value, they do indicate that threatened species are better represented in *ex situ* collections than non-threatened species. Given that threatened species are harder to find and often much more difficult to cultivate and manage, this is compelling evidence that botanic gardens are taking plant conservation seriously, and stepping up their efforts to avoid plant species extinctions.

5.2 Living Collections

For the purposes of this discussion, we are defining living collections in botanic gardens as *growing* plants in botanic garden landscapes, glasshouses, nurseries and laboratories – as opposed to seeds, spores and other propagules that, although living, are not growing.

5.2.1 Living Collections as a Source of Material for *Ex Situ* Species Conservation, Reintroduction and Translocation

As indicated above, latest estimates suggest that around 70 per cent of threatened European plant species are held in *ex situ* living collections. However, one-third of these are single collections, meaning that they are not genetically diverse, a topic that will be covered later in this chapter. However, it is worth looking at the conservation value of genetically homogenous living collections in a little more detail here. Infra-specific diversity is important for resilience because useful genetic traits, such as disease resistance and abiotic tolerance, are not universally present in populations. They may occur only in subsets of populations or in individuals and therefore it is important to sample across populations when plants are collected for conservation purposes such as species reintroduction or restoration. However, the relative importance of infra-specific genetic diversity varies according to factors such as the reproductive strategy of a plant (e.g. out-crosser versus selfer), gene flow between populations and the adaptability of a plant.

In the conservation context, there is also the question of whether self-sustaining populations in the landscape are a realistic outcome. In many cases, plant species in 'the wild' are ecologically isolated from their pollinators or dispersers due to the fragmentation of the habitats and landscapes in which they occur. Unless these commensal species can be conserved and are self-sustaining alongside the plant species with which they are symbionts, dependent plant populations will not persist. There are already ample examples of plant species that are extinct in the wild but which persist in horticulture in botanic gardens or elsewhere. For example, the North American tree species *Franklinia alatamaha*, has been extinct in the wild since the early nineteenth century but it thrives in gardens. Similarly, Wood's cycad, *Encephalartos woodii*, which originates in South Africa, has been maintained from cuttings from a single plant for the past 120 years in botanic gardens around the world. Neither of these species is likely to survive without human management and assisted reproduction. However, there are other species that have been brought back from the brink of extinction and reintroduced to the wild – including some from very narrow genetic stock. One of these is *Erica verticillata*, a South African species extinct in the wild for over 100 years but reintroduced to Rondevlei and Tokai nature reserves near Cape Town from cuttings from Kew and Pretoria.

Species reintroductions and translocations from botanic garden-generated material are commonplace these days, with hundreds of species reintroduced to the wild or translocated, as highlighted in Chapter 6 of this book. Guidelines on plant species reintroductions have been produced by the Center for Plant Conservation (Maschinski *et al.* 2012),[3] and translocation guidelines for rare and threatened plants are available from the Australian Network for Plant Conservation.[4]

5.2.2 Living Collections as a Research Resource

The value of living collections in botanic gardens is not just restricted to their potential for reintroduction to the wild. Living collections are a resource for researchers studying how to conserve or manage species in their natural habitats, or for cultivation in the man-managed landscape. For example, oak trees (*Quercus* spp.) at the Royal Botanic Gardens, Kew, have been used to study the life cycle and control of oak processionary moth which not only affects trees but is a health hazard to people. Similarly, Kew's ash (*Fraxinus*) collection has been used to test for natural resistance to the fungal disease ash dieback (*Hymenoscyphus fraxineus*) (Kirkham, pers. comm.). Although, in general, living plant collections are underutilised for this kind of research, some botanic gardens are developing their role in understanding and preventing plant pests and diseases further through the International Plant Sentinel Network (IPSN) – a botanic garden-centred early warning system for emerging plant pests and diseases (see Chapter 3).

Living collections in botanic gardens can also be used to define the abiotic tolerances of plants, which is useful for predicting how our common (or rare) plant species will adapt to the changing climate (Sax et al., 2013). Botanic garden collections are also useful for predicting the invasiveness of species and whether a species might become naturalised or even domesticated (Early and Sax, 2014). In the case of domestication, Malawi's national tree, the Mulanje cedar (*Widdringtonia whytei*), is known from botanic garden data to grow well at substantially lower altitudes than those of its 'natural range' on Mount Mulanje. It is highly likely that the cedar's current range is a refugium and not optimal for the species. As this book goes to press, trials supported by the UK government's Darwin Initiative are being set up for the cedar across Malawi at lower altitudes based on observations in gardens in New Zealand, Kenya and Indonesia.

The role of living collections in plant domestication goes all the way back to the age of economic botany and was the *raison d'être* for many botanic gardens. Gardens such as David Fairchild's at the Kampong (see Case Study 5.1) in Miami, Florida, were the first places where newly imported plants were established so that they could be studied, their performance assessed and where they could be quarantined if necessary. This is still a key role of botanic gardens today, particularly supporting the horticultural industry with new introductions that may or may not find their way into commercial horticulture.

5.3 Seed Banks

Seed banks offer the opportunity to conserve large amounts of plant diversity cheaply and effectively for decades and, in some cases, centuries. This technology has mainly been applied over the past 60 years to conserving the diversity within the relatively few domesticated (crop) species, thereby making it available for varietal improvement. However, over the past three decades, there has been increasing interest in the use of

> **Case Study 5.1** *The Kampong*
>
> The Kampong in Coconut Grove, Florida was originally the home of David Fairchild who was instrumental in the development of agriculture and horticulture in the United States. Today, The Kampong is part of the National Tropical Botanical Garden, a non-profit organisation headquartered on Kauai, Hawaii. Although The Kampong is only a small garden of 11 acres, it contains over 1,400 living collections, many of economic value. The collection began as a personal collection motivated by Dr Fairchild's love for, and scientific interest in, edible, ornamental and ethnobotanical plants. Today, the garden is renowned for its historic collections or rare and unusual varieties of tropical and subtropical fruits, palms and flowering trees, shrubs and vines (Schokman, 2012). It was on this site, while Dr Fairchild was Chief of the Foreign Seed and Plant Introduction Section of the United States Department of Agriculture (1897–1928), that he introduced, tested and grew many tropical plants – primarily species that were new to the US, collected on his various plant explorations throughout the world. During his 30-year tenure, almost 75,000 plants were introduced and grown in the US. He continued this work after his retirement, and the collection was further expanded by the garden's next owner, Dr Catherine Sweeney and ultimately by the NTBG, the garden's current proprietor. Today, the plants at The Kampong represent a unique collection with a distinctive history and identity.

this technology to conserve non-domesticated (wild) species. Such collections are a resource for habitat restoration and afforestation by enabling species to be put back where lost or in creating new plant communities adapted to future environmental conditions. They are also a huge untapped resource for research and new technology, not least in agriculture, forestry and horticulture.

5.3.1 Seed-Banking Protocols: The Basics

Seed banking is cheapest and most effective for plant species that produce orthodox seeds, i.e. seeds that are desiccation tolerant and can be dried down to a low moisture content before freezing. At least three-quarters of the world's plant species produce orthodox seeds. For the remainder – termed 'recalcitrant' or 'intermediate' in their seed behaviour – other techniques (covered later) are required for their long-term conservation, *ex situ*. Orthodox seeds are typically small with hard seed coats, and usually have a dormant state that is only broken when the right environmental conditions, chemical trigger or physical damage to the seed coat occurs. Recalcitrant seeds, on the other hand, are usually large with thin seed coats and have a short or no dormant phase (Tweddle *et al.*, 2003). Following standard seed-banking protocols (Smith *et al.*, 2003), seeds are dried down to typically 6–7 per cent moisture content in a drying room or desiccator at 15 °C and 15 per cent relative humidity. Once dry, the seeds are packaged in aluminium foil or glass containers and stored at −20 °C. Under these conditions, most orthodox species will live for decades and even hundreds of years without significant loss of viability (Walters *et al.*, 2005). However, a small number of orthodox species are short-lived in these conditions (Probert *et al.*, 2009), and ultra-low temperatures may be used for their long-term storage.

Standard seed-banking protocol requires that seed viability is tested before seeds are banked so that a baseline for seed viability is created against which survival can be measured. To do this viability testing requires the development of optimal germination protocols, which can be challenging for wild species, particularly those with complex dormancy mechanisms. Kew's Millennium Seed Bank uses a computer-based germination predictor tool to predict the optimal conditions for germination by taking into account the climatic conditions where the species was collected and what is known about the nature of a species' dormancy based on its seed morphology and autecology.[5] Once an optimal germination protocol has been developed, seeds can be banked and then re-tested for viability periodically – typically every 10 years – by removing a sample of seeds from the freezer and germinating them. A useful reference for optimal germination protocols, seed behaviour and other traits is Kew's Seed Information Database, a compendium of seed information.[6]

5.3.2 The Purpose of Seed Banks

Virtually all seed banks have two main purposes:

1. To store seed material for (long-term) conservation purposes as a backup or insurance policy should that material be lost in the wild or in the landscape.
2. To supply seeds for research and use.

The conservation role of seed banks is well understood but their role as a supplier of material for use is not. The common public perception of seed banks is that they are 'doomsday vaults', there only in case of disaster. Seed banks do fulfil this role as we have seen in the case of the Svalbard Seed Vault maintained by the Global Crop Diversity Trust which has recently returned duplicate collections of seeds destroyed in the Syrian conflict to the International Centre for Agricultural Research in Dry Areas (ICARDA). However, the primary purpose of seed banks is to supply seeds for use. This is as true in wild seed banking as it is in the crop sector.

The botanic garden community includes the largest, most sophisticated seed banks in the world. Kew's Millennium Seed Bank (see Case Study 5.2) is the most diverse seed bank on the planet. The Royal

---

Case Study 5.2 *Kew's Millennium Seed Bank Partnership*

Seed banking of wild species was pioneered by, amongst others, RBG Kew's Seed Conservation Section at Wakehurst Place in Sussex. This outpost of Kew's Jodrell Laboratory was established in the 1970s, and whereas in the past seeds had been banked by many botanic gardens for use in horticulture and research, it was then recognised by the Kew group that banking seed for conservation purposes was an urgent need. In the mid 1990s, RBG Kew applied for lottery funding to develop their modest seed vault (located in the disused chapel of a sixteenth-century building) into the largest seed bank in the world. The Millennium Seed Bank cost £18 million to build and was completed in 2000. Making the argument that, just as a library isn't a library without books, a seed bank isn't a seed bank without seeds, Kew raised a further £50 million to develop the 10-year Millennium Seed Bank Project (MSBP) which had two major aims:

1. to collect and bank seeds from 10 per cent (24,200) of the world's plant species, prioritising rare, threatened and useful species; and
2. to build seed conservation capacity worldwide through partnership, training and information exchange.

Working in partnership with over 120 botanic gardens, forestry institutes and agricultural organisations in 50 countries, the MSBP trained over 2000 people in seed conservation techniques, set up and improved seed banks in its partner countries and achieved its 10 per cent target on time and budget.

During that first 10-year phase, it became clear that simply banking the seed was not meeting the needs of many of the MSBP partner countries. As well as the need to conserve species, there was great demand to use seed for development and conservation purposes. To this end, in 2010, the Millennium Seed Bank Project became the Millennium Seed Bank Partnership working towards a new 10-year plan that aimed to:

1. bank the seed from 25 per cent of the world's plant species by 2020, prioritising rare, threatened and useful species; and
2. enable the use of seed collections for innovation, adaptation and resilience in agriculture, forestry, horticulture and habitat restoration.

Today, the MSB Partnership is operating in over 80 countries, and its portfolio includes projects addressing food security, water scarcity, human health, energy, habitat restoration and species reintroductions.[7]

Botanic Garden, Sydney's PlantBank, completed in 2014, is the largest seed bank in the southern hemisphere. The Kunming Institute of Botany's Gene Bank of Wild Species (GBOWS) has already collected and conserved over 10,000 species – a third of China's native flora. Both PlantBank and GBOWS grew out of the Millennium Seed Bank Partnership, and many smaller seed banks have also been established in botanic gardens as a result of the MSBP. The Global Seed Conservation Challenge of BGCI (see Case Study 5.3) comprises around 180 botanic gardens with wild species seed banks.

5.4 Exceptional Species

Exceptional species is a designation given to those species for which traditional seed banking conditions, i.e. drying at 15 per cent relative humidity and freezing at -18 to $-25\,°C$, are not workable. These conditions are readily applied to the majority of species of seed plants, as noted in Section 5.3. For the remainder, there are several reasons for their exceptional status. Species with seeds that are intolerant of desiccation, known as *recalcitrant* species, will be killed by the drying required for the banking

Case Study 5.3 *BGCI's Global Seed Conservation Challenge*

BGCI's Global Seed Conservation Challenge (GSCC)

Target 8 of the Global Strategy for Plant Conservation (GSPC) calls for '75 per cent of threatened plant species in *ex situ* collections, preferably in the country of origin and at least 20 per cent available for recovery and restoration programmes'. The Convention on Biological Diversity's mid-term review of the GSPC (Sharrock *et al.*, 2014) highlighted that more needs to be done if the 2020 targets are to be achieved.

Ex situ collections of plant species act as an insurance against extinction in the wild while being available for research, reintroduction and restoration. Botanic gardens are the main institutions involved in *ex situ* conservation of threatened species with nearly one-third of all known plants grown in botanic gardens. More than 400 botanic gardens worldwide are banking seed for conservation.

The GSCC seeks to promote and support *ex situ* seed conservation in the botanic garden community. All gardens that actively contribute to *ex situ* seed conservation are making progress towards the ambitious objectives of Target 8 and are therefore participating in the GSCC.

The challenge aims to:

- Engage more botanic gardens in seed banking, working 'outside the garden walls' to bring threatened species that are not already conserved into *ex situ* collections.
- Strengthen networks to help botanic gardens share experiences and resources in seed banking.
- Establish a seed collecting 'hub' at BGCI which will provide a 'one-stop-shop' for seed banking information and training resources.
- Provide training and build capacity to support seed collecting and raise seed banking standards.
- Award prizes for seed conservation at the Global Botanic Garden Congresses.

Botanic gardens participate by uploading their seed collection data to PlantSearch,[8] sending BGCI seed banking stories and entering BGCI's seed banking competition.[9]

> BGCI provides training, technical resources and financial support for seed conservation. It also carries out gap analysis studies, and provides information on which species are not represented in *ex situ* collections, enabling gardens to prioritise species for seed collection. Finally, BGCI highlights stories and news from botanic gardens about their seed conservation work, and awards prizes for achievement in seed conservation at its global congresses.[10]
>
> To date, BGCI's Global Seed Conservation Challenge has 140 member gardens.

process itself. Additionally, some species, due to reproductive failure or extreme rarity, produce few or no seeds, and thus lack sufficient seeds for banking. Finally, there are species with various behaviours that similarly preclude their storage in standard seed banks. Some seeds, termed *intermediate*, while desiccation tolerant, are sensitive to cold temperatures. Other seeds, often termed *short-lived*, as well as the spores of pteridophytes, tolerate desiccation, but appear to be somewhat intermediate between recalcitrant and orthodox, in that they tolerate the conditions of traditional seed banks, but do not maintain viability as required for long-term *ex situ* conservation, and they will be included in this discussion, as well.

There are several challenges to the *ex situ* conservation of exceptional species that are not shared to a large degree by non-exceptional species. These have been outlined in a statement issued as the result of a workshop on conserving exceptional species convened by Botanic Gardens Conservation International and the Cincinnati Zoo and Botanical Garden, as part of the fifth Global Botanic Gardens Congress in Dunedin, New Zealand, in October, 2013 (see Box 5.1). They centre on the need for identifying which endangered species are exceptional, for better scientific understanding of the species, and for meeting the challenge of additional costs of conserving exceptional species and facilitating global communication and coordinating efforts to that end.

Identifying exceptional endangered species will require collective evaluation of information on species of conservation priority (e.g. the IUCN Red List and other regional or national red lists), on seed storage requirements and availability of information (e.g. Kew's Seed Information

> Box 5.1 *Conserving Threatened Exceptional Plant Species: Statement of Need*
>
> NEED 1: ADDRESS INFORMATION CHALLENGES. We have incomplete information on which threatened plant species are exceptional, and which researchers and practitioners are currently working with threatened exceptional species. These are currently the most significant information barriers to effective conservation of threatened exceptional plant species, and therefore a priority for action.
>
> NEED 2: IDENTIFY RESEARCH PRIORITIES. The biology of most threatened exceptional plant species is generally poorly understood and often species-specific. *In vitro* propagation and cryopreservation are currently the primary techniques to conserve threatened exceptional species *ex situ* and to produce propagules for reintroduction efforts, but specialised facilities and expertise are often required to develop protocols for these techniques, and they are often species-specific. Even under the best of circumstances, protocol development is often expensive, time consuming and unpredictable.
>
> NEED 3: ADDRESS FUNDING, COMMUNICATION AND COORDINATION CHALLENGES. Conserving threatened exceptional plant species *ex situ* is more costly than traditional seed banking. In general, cryopreservation and *in vitro* propagation costs are difficult to quantify or standardise, are species-specific, and results are unpredictable. Additional funding for research, outreach, and more effective communication and coordination among the global community working with these species, is needed. Potential avenues to secure funding to support research, outreach and coordination include targeting major foundations, the corporate sector and industry leaders.

Database, CPC Species Profiles), and personal communications from researchers. Bringing all of this information together into a global list of exceptional species will be a challenge, although a start has been made in the publication of a list of North American exceptional species, which may form the basis of more extended, global efforts in the future (Wallace, 2015).

The other challenges of conserving exceptional species centre on the fact that the methods required have significantly greater costs than traditional seed banking (Pence, 2011), due to the need for methods requiring more labour and supplies than seed banking. Although a few standard methods appear to be generally adaptable to many species, there is often a period of protocol development needed for applying methods to a new species. Continuing research on the biology of these species and their response to these protocols is required, as anything that can make cryopreservation, as well as the *in vitro* methods often needed to support it, more efficient will reduce costs and ensure that more can be done for exceptional species with limited resources.

5.4.1 The Science and Technology of Exceptional Species Conservation

5.4.1.1 Seed Plants

A variety of methods are available for approaching the *ex situ* conservation of exceptional species, all of which are, to a greater or lesser degree, more labour intensive and resource requiring than traditional seed banking. Exceptional species can be preserved in living collections within botanic gardens, as discussed earlier in this chapter, but this section will focus on propagules that provide an alternative when seeds are not adaptable or available for seed banking. There is a hierarchy of methods available for the *ex situ* conservation of exceptional species, which can be ordered according to the amount of resources generally needed (see Box 5.2). Methods also differ in the amount of genetic diversity captured and the resources needed for recovery. Evaluating all of these factors can help in prioritising species and selecting methods for conserving them. The two primary technologies involved in these methods are cryopreservation, or storage in liquid nitrogen (LN), and, in some cases, *in vitro* culture (tissue culture), or growth of the tissue aseptically on an artificial nutrient medium. Both require some specialised equipment and expertise.

If desiccation tolerant seeds are available, seed banking should be the first choice. The exception to this would be those species with short-lived seeds, which may require $-80\ °C$ or LN for adequate long-term survival. When seeds are available but are recalcitrant, the next most efficient option is embryo axis cryopreservation (Normah and Makeen, 2008), which is possible in those species where the embryo axis is sufficiently desiccation tolerant, even if the entire seed is less tolerant. This can capture a range of genetic diversity similar to seed banking. Key steps in this procedure are isolating intact axes

Box 5.2 *Selecting* Ex Situ *Conservation Methods for Seed Plants*

			Storage Method	Genetic Diversity	Recovery Method
1.	(a) Seeds available,	go to 2			
	(b) Seeds not available,	go to 5			
2.	(a) Seeds desiccation tolerant,	go to 3			
	(b) Seeds are desiccation sensitive,	go to 4			
3.	(a) Seeds not short lived or intermediate		Seed bank	High	Germination
	(b) Seeds short lived or intermediate		Cryopreserve	High	Germination
4.	(a) Embryo axis desiccation tolerant		Cryopreserve	High	Micropropagation Germination
	(b) Embryo axis desiccation sensitive,	go to 5			
5.	(a) Dormant buds are available,	go to 6			
	(b) Dormant buds are not available		*In vitro*	Low–Moderate	Micropropagation
6.	(a) Dormant buds freezing tolerant		Cryopreserve	Low–Moderate	Graft, root, micropropagation
	(b) Dormant buds not freezing tolerant		*In vitro*	Low–Moderate	Micropropagation

from the seeds, drying and/or applying chemical cryoprotection, and then cryopreserving the axes in LN. For recovery, the embryos are surface sterilised and cultured *in vitro* for germination, in order to provide nutrients to the embryo, normally supplied by cotyledons or endosperm. For some species, research is needed to better understand the conditions required for *in vitro* germination (Xia et al., 2014).

A further possibility is the use of dormant buds (Sakai and Nishiyama, 1978; Volk et al., 2009). The amount of genetic diversity captured depends on the ability to collect buds from genetically distinct individuals in the field, as all material generated from each bud will be clonal. Key steps in this procedure are the collection and cutting of the buds, and additional drying or cryoprotection, if needed. Recovery methods have included *in vitro* culture, grafting and direct rooting of the buds, depending on the species. This procedure will likely have wider application, but may be limited to woody, cold-climate species with buds that overwinter in the dormant state.

When these methods are not applicable, shoot tips or somatic embryos (Dereuddre *et al.*, 1991; Reed, 2008) can provide an additional method for *ex situ* conservation. Since these methods require the initiation of an *in vitro* culture as a source of material for cryostorage, as well as for recovery, they are generally considered the most costly of the seed banking alternatives. Key steps in this method are surface sterilising shoot tips or other tissues and culturing on a medium that can stimulate the outgrowth of lateral shoots or the production of somatic embryos. These cultures can be maintained by subculture every 4–12 weeks. For the cultures to be useful for conservation, protocols must also be developed for recovering whole plants from these, either by rooting of shoots or stimulating the growth of the embryos into plants.

In vitro cultures can be used in two ways for *ex situ* conservation. They can be maintained indefinitely as an *in vitro collection* by periodic subcultures and used as a source of material for research, restoration and as a back up for critically endangered species in the wild. Such a collection has been created and maintained at the Micropropagation Laboratory at the Lyon Arboretum (US) to aid in the conservation of critically endangered Hawaiian taxa. *In vitro* collections are resource intensive, but can conserve more species diversity in less space than traditional living collections. Such collections do require a continuous input of labour and resources (such as growth media and containers) and growth room space, and hold the potential for genetic changes and selection over the long term. These disadvantages can sometimes be minimised by developing protocols for *slow growth* culture, either by lowering the nutrients available or by lowering the temperature, or both (Keller *et al.*, 2006). This has proven effective in lengthening the time between subcultures in a number of species, thereby reducing the labour and resources needed for maintenance.

The second approach is to use *in vitro* cultures as source material for cryopreservation. In this case, either tiny, 1 mm shoot apices or somatic embryos are used. There are a few basic methods that have been used for tissue cryopreservation, including slow freezing (two-step freezing), requiring relatively expensive equipment for slowly reducing the temperature (Reed and Uchendu, 2008); encapsulation dehydration, a method for cryoprotection with sucrose and drying (Fabre and Dereuddre, 1990); several forms of vitrification, in which a concentrated solution of chemical cryoprotectants is used, including vitrification, encapsulation vitrification, and droplet vitrification (Sakai *et al.*, 1990; Matsumoto *et al.*, 1995; Panis *et al.*, 2005). Once thawed, the shoot tips or embryos must be recovered *in vitro* and the protocols developed for plant production can be used to make plants available for restoration and research.

Because *in vitro* propagation is clonal, the genetic diversity captured in tissue culture collections or cryopreserved tissue collections will depend on the number of genetically distinct lines initiated and maintained or cryopreserved, and the limiting factor may be the resources needed to maintain each one of those lines separately. Cryopreservation offers the possibility of banking such lines, so that genetic diversity is preserved, without the need for continuous subculture.

5.4.1.2 *Spore Plants*

Just as seed banking is the most efficient method for maintaining the genetic diversity of seed plants *ex situ*, spore banking is a similarly effective corollary method for preserving the genetic diversity of some spore-bearing species. Fern spore banking has received less attention than seed banking, but botanic gardens have been in the forefront of this technology. One of the oldest efforts in spore banking has been at the Royal Botanic Garden Edinburgh (Scotland, UK), with other collections at the Bank of Fern Spores at the University of Valencia (Spain) and at CREW at the Cincinnati Zoo and Botanical Garden (US).

Pteridophyte spores are classified as green (chlorophyllous) and non-green (non-chlorophyllous). Non-green spores are generally longer-lived than green spores under ambient conditions and were originally thought to be analogous to orthodox seeds, with green spores thought to be like recalcitrant, or desiccation sensitive, seeds. However, it appears that green spores are also desiccation tolerant, although they decline in viability more rapidly after maturity than non-green spores, and thus, must be dried and stored while they are very fresh (Pence, 2000a; Li and Shi, 2015). In addition, both green and non-green spores appear to lose viability more quickly than expected when maintained dry at $-25\,°C$, and storage at $-80\,°C$ or in LN is recommended for preserving viability in the long term (Ballesteros *et al.*, 2011, 2012). It has been concluded that fern spores possess a behaviour, as do some seeds, that is intermediate between orthodox and recalcitrant, or they may possess a response that has not yet been fully characterised (Ballesteros *et al.*, 2012).

While pteridophyte spores have received less attention than seeds, there has been even less work on the conservation of bryophyte spores.

However, reports of desiccation tolerance in spores, as well as of diaspores and brood cells, of at least some bryophyte species strongly suggest that such spores, like desiccation tolerant fern spores, would be good candidates for long-term cryopreservation (Duckett *et al.*, 1993; Goode *et al.*, 1993). Work is needed to determine whether they resemble fern spores in their intermediate behaviour.

Another option for both bryophyte and fern conservation is the use of gametophytes. Gametophytes of both groups can be grown *in vitro*, and *in vitro* bryophyte collections are becoming important conservation tools in the UK, Romania, Serbia and elsewhere (Rowntree and Ramsay, 2005; Cogălniceanu, 2014; Sabovljević *et al.*, 2014). *In vitro* gametophytes of both ferns and bryophytes have also been successfully cryopreserved using several methods (Christiansen 1998; Pence, 2000b; Rowntree *et al.*, 2011), and these procedures build on the natural adaptations of many gametophytes to survive desiccation and freezing (Oliver *et al.*, 2005; La Farge *et al.*, 2013), providing opportunities for efficient *ex situ* conservation. A six-year effort at the Royal Botanic Gardens, Kew, focused on the *in vitro* culture and cryobanking of endangered bryophytes of the UK (Rowntree *et al.*, 2011), and other reports support the use of these methods for the conservation of endangered bryophyte taxa (Sabovljević *et al.*, 2012; Vujičić *et al.*, 2012).

Algae are less represented as areas of focus in botanic gardens, although some, such as the New York Botanical Garden, have well-developed programmes dealing with algal systematics. Work at universities and marine institutes, however, has resulted in the development of methods for maintaining some algae, particularly some microalgae, in LN storage. These efforts have been fuelled by an interest in facilitating the long-term maintenance of important collections of microalgae and cyanobacteria (Brand and Diller, 2004), as well as for conservation of algal biodiversity, as in the European-wide effort, COBRA (Day *et al.*, 2005). One of the oldest cryostored collections (40 years), the Culture Collection for Algae and Protozoa at the Scottish Marine Institute, is preserving microalgae (Childs *et al.*, 2015). There has been less work on macroalgae, although success with cryopreserving vegetative and reproductive tissues of several species has demonstrated the potential in this area, both for maintaining important research collections and for biodiversity conservation (Heesch *et al.*, 2012; Green and Neefus, 2014; Lee and Nam, 2016). Desiccation tolerance may also differ between life forms within algal species and this could inform conservation strategies (Luxoro and Santelices, 1989). Botanic gardens with expertise in algae biodiversity might partner with centres of expertise in algae cryopreservation to build long-term collections of wild, non-model species for research and restoration in the future.

5.4.2 Capacity for Dealing with Exceptional Species

Botanic gardens have an important role to play in the conservation of endangered exceptional species, since they often are deeply involved in the identification and monitoring of the rare species of their regions. However, currently very few gardens have programmes and infrastructure to do so, although the number is slowly increasing. Three of the oldest programmes that include both *in vitro* and cryopreservation work include the Royal Botanic Gardens, Kew in the UK, Kings Park and Botanic Garden in Australia, and CREW at the Cincinnati Zoo and Botanical Garden in the US (see Case Study 5.4). More recently, a number of

---

Case Study 5.4 *The Center for Conservation and Research of Endangered Wildlife, Cincinnati Zoo and Botanical Garden*

Founded in 1980, at Cincinnati Zoo and Botanical Garden (CZBG), with a focus on using technologies for animal conservation, the Center for Conservation and Research of Endangered Wildlife (CREW) added a Plant Research Division (PRD) in 1987, interpreting 'wildlife' to include both plants and animals in need of conservation. The focus of the PRD is to use biotechnologies for the propagation and preservation of endangered plants when traditional methods are not adequate. It is the home of CREW's Exceptional Plant Signature Project. With a focus on using *in vitro* methods and cryopreservation, CREW partners with institutions across the US to develop propagation and preservation protocols for US endangered plants and also conducts basic research on exceptional species with the goal of improving growth and making protocols more efficient. In partnership with gardens within the Center for Plant Conservation network, several species have been produced for restoration projects, including the autumn buttercup (*Ranunculus aestivalis*) and the Cumberland sandwort (*Minuartia cumberlandensis*). The *In Vitro* Collection holds over 50 species in active culture, while the Frozen Garden within CREW's CryoBioBank holds over 2,000 samples of seeds, spores, gametophytes, pollen, embryos and shoot tips from over 150 species, and samples have been recovered after up to 23

> years in cryostorage. CREW's work, as well as plants resulting from propagation and cryostorage, are interpreted for CZBG visitors, which number over 1.5 million per year.

botanic gardens globally are adding research programmes, some of which are including tissue culture labs and cryostorage capacity. The number of threatened exceptional species is currently unknown, but a recent estimate of North American threatened species identified 22 per cent as likely exceptional (Wallace, 2015). Tropical moist forests may harbour more (Tweddle *et al.*, 2003; Pritchard *et al.*, 2014), with the actual global number in the thousands. Considering the additional effort and expense in comparison with seed banking, this is a daunting challenge for the conservation and botanic garden community.

Such an effort will require not only funding but collaborations. Botanic gardens with capacity could be centres of expertise, providing services for other gardens. Additionally, government or academic labs, which may have infrastructure, as well as student labour, could similarly partner with botanic gardens that possess access to and knowledge of rare exceptional taxa. The seeds for such collaborations were sown at the exceptional species workshop in Dunedin, in 2013, with the formation of the Exceptional Plant Species Advisory Group (EPSAG), and further efforts are needed to co-ordinate an expanded focus on exceptional species conservation.

5.5 The Strengths and Weaknesses of *Ex Situ* Collections for Conservation Purposes

The primary goal of *ex situ* collections is to maintain a representation of the species as a source of material for restoration, should the species be lost in the wild, and this should be done as effectively and efficiently as possible. As an overall approach, the maintenance of *ex situ* collections is a powerful tool supporting conservation, with the insurance of rare taxa held in botanic gardens worldwide an invaluable resource for the future. Individually, each conservation method as outlined previously has particular strengths and limitations and these should be considered when undertaking *ex situ* conservation.

Protection of the collection is of first importance, since these collections are meant to hedge against extinction in the wild for decades, and each approach has its own vulnerabilities. Living collections can be subject to disease and damage through natural disaster, as well as theft, and some botanic gardens have supplied duplicate samples to other institutions to maintain a backup to their collection. *In vitro* collections, which are maintained in protected laboratories free from apparent disease, must be routinely subcultured, which holds the potential of introducing contamination which can overwhelm and kill the tissues. Tissues that are cryopreserved as aseptic samples are free from infection while stored, but in both *in vitro* and cryopreserved collections, damage to the supporting infrastructure, either through natural disaster or system interruption or failure, could place *ex situ* samples at risk.

For effective conservation of a species, the collection should include enough genetic representation to replicate the diversity of the species in the wild in a restoration programme. Seed and spore banking can effectively store the genetic diversity of a species when collected from an adequate representation of populations and individuals and handled appropriately, as described in several available sets of collecting guidelines (Guerrant, *et al.*, 2004; FAO, 2014). Storage of isolated embryo axes, taken from seeds, could provide a similar function with proper collecting strategies. Dormant buds, shoot tips and somatic embryos, however, are clonally reproduced and sufficient genotypes must be represented to produce an effective collection, requiring a significant input of labour and resources.

With all of these methods there is the possibility – indeed the likelihood – of selection occurring in the handling, storage or growth of the species. Documenting the initial genetic diversity of the collection can help document any changes that may occur over time. On the other hand, tissues maintained *in vitro* are at risk of spontaneous genetic changes, or somaclonal variation, which has been documented for a number of species (Larkin and Scowcroft, 1981). While plants in living collections appear to be more stable than those in culture, they may also experience selection for fitness to grow in a garden setting, which may be different from their normal evolutionary course. All forms of *ex situ* collections should have provenance documentation, which can then be connected with any measured genetic diversity.

The different approaches to *ex situ* collections also vary significantly in cost. Costs of living collections can be quite high, while the genetic diversity maintained may be quite low, due to space and facilities limitations (Li and Pritchard, 2009). *In vitro* collections, which require periodic maintenance, have ongoing facilities and labour costs, while cryopreserved collections require significant facilities and labour initially, but once banked, their long-term maintenance is comparable to those of traditional seed banking (Reed *et al.*, 2004; Li and Pritchard, 2009; Pence, 2011).

5.6 Challenges Associated with Maximising the Value of *Ex Situ* Collections for Conservation

There are significant challenges associated with maximising the value and use of *ex situ* collections for plant conservation. Most of these challenges are not technical but are cultural and political. In Chapter 10 we cover some of these issues, and the changes that botanic gardens need to make from within – for example, directing more resources towards plant conservation, prioritising efforts on rare and threatened species, and working with other sectors. In this section, we will look at how botanic gardens are perceived from the outside, their credibility, potential niche and how they can best address current knowledge gaps.

5.6.1 Defining the Niche for Botanic Gardens in Plant Conservation Practice

Botanic gardens have always been useful as sources of data for plant conservation *in situ*. Plant identification tools such as floras, keys and field guides, locality data from herbarium sheets, information on autecology from plant physiology studies and so on are extremely useful to field botanists and conservationists. However, botanic gardens are not as readily recognised as sources of plant material and knowledge about how to grow plants. The concept of *ex situ* plant conservation as a 'backup' or insurance policy against extinction in the wild is well understood by most conservation practitioners. However, in this context, *ex situ* conservation is often seen as a last resort. The necessity for *ex situ* conservation, as part of the armoury of techniques required for integrated plant conservation, has sometimes been difficult to make to *in situ* conservation practitioners in the past. However, it is estimated that humans have modified more than 50 per cent of the world's land surface (Hooke *et al.*, 2012), with approximately 40 per cent given over to agriculture and livestock management. For plants with natural distributions that fall within these transformed areas, *ex situ* conservation or active human management may be the only way they can survive. Even in national parks and wilderness areas not significantly altered or actively managed by people, plant populations may be vulnerable – particularly to invasive species, pests, diseases and a changing climate.

In spite of this very clear and well-documented trend of human-driven land transformation, most *in situ* conservation efforts are directed at saving what is left of wild habitats rather than restoring what has been degraded. Given that resources are limited, this approach makes sense. However, with only around 15 per cent of the world's land surface area receiving legal protection in the form of national parks, reserves and protected areas, those species occupying the rest are rapidly being driven to extinction. This creates both a challenge and an opportunity for botanic gardens. As indicated earlier, the botanic garden community conserves and manages a far greater range of plant diversity than any other sector. Moreover, it does so largely in human-managed, transformed landscapes. We argue that botanic gardens have a major role to play in preventing plant species extinctions through integrated plant conservation action. This argument is based on the following assumptions:

- There is no technical reason why any plant species should become extinct. Given the array of *ex situ* and *in situ* conservation techniques employed by the botanic garden community (seed banking, cultivation, tissue culture, cryopreservation, assisted migration, species recovery, ecological restoration, etc.) we should be able to avoid species extinctions.
- As a professional community, botanic gardens possess a unique set of skills that encompass finding, identifying, collecting, conserving and growing plant diversity across the entire taxonomic spectrum.

While these assumptions are readily defensible, and there is plenty of evidence to back them up, few gardens see prevention of plant species extinctions as their main purpose. Many – perhaps most – gardens do not recognise this as their purpose at all. At the same time, ironically, many gardens struggle to make the case to funders and policymakers that they have an essential role to play in society – a specific niche that no other sector can fill.

Botanic gardens are complex organisations, fulfilling many objectives that frequently compete with each other and with other sectors. As visitor attractions, botanic gardens compete with museums, zoos, theme parks, public gardens, historic sites, cinemas, theatres and a myriad of other recreational facilities. As educational centres, they compete with museums, zoos, a wide range of educational charities, universities and even schools. As horticultural attractions, botanic gardens are in competition with municipal parks, public gardens, historic gardens, private gardens and even nurseries and garden centres. In plant research, our rivals include universities, agricultural colleges, forestry research institutes and commercial companies. Finally, in plant conservation, we compete with a vast array of conservation NGOs and charities conserving charismatic species such as tigers, pandas and golden eagles.

It could be argued that the only thing that botanic gardens do *uniquely* well is to collect, document, describe, grow, conserve and manage plant diversity across the taxonomic array. Given the current Anthropocene extinction crisis, it seems entirely logical that we should scale these activities up urgently. One of the issues we need to consider in scaling up *ex situ* conservation activities at a global level is the legal framework for exchange of plant material.

5.6.2 Access and Benefit Sharing (ABS)

The tenets and obligations of the Convention on Biological Diversity (CBD) and the subsequent International Treaty on Plant Genetic Resources in Food and Agriculture (ITPGRFA) govern international collection and exchange of plant material. As well as being a global framework to catalyse conservation action, the CBD, in particular, has created some obstacles to effective species conservation. Professor Sir Peter Crane, a former director of the Royal Botanic Gardens, Kew, whose tenure was characterised by significant financial and intellectual investment into making the CBD work, eloquently articulates some of these obstacles. He says:

> ... while the text of the convention acknowledged that the conservation of biological diversity 'is a common concern of human kind,' it nevertheless enshrined into international law for the first time the principle that biological resources within the borders of individual countries are a national patrimony. At one level, this simply reaffirmed practices already implemented in many countries, where the use of animals and plants living within their borders was subject to national laws, but when the issue was raised in international negotiations leading to the CBD, the question of under what terms biological resources should be shared internationally was given new prominence. Inevitably, it inspired new nationalistic sensitivities and further complicated the political landscape ... It was a seemingly sensible approach, but it had at its heart one fundamental consequence: no longer were the plants and animals bequeathed to all of us by 4.6 billion years of planetary evolution part of our common human patrimony; instead they became the proprietary interests of nations. Trees, birds, flowers, and all other kinds of organisms from insects to bacteria, and the genetic

material they contained, were taken as property by the people living inside the borders of individual countries.

He goes on to say:

One unintended consequence is that by limiting access through complex permitting regulations the CBD actually helped stifle commercially oriented work that could potentially generate revenue. Another is that many countries greatly restrict scientific access to their native plants and animals, even for non-commercial work in collaboration with in-country scientists that helps support animal and plant conservation. (Crane, 2013)

Botanic gardens, in pursuing their plant conservation and research activities, have to try to navigate these obstacles. This is particularly the case where international cooperation is involved. Here the legal landscape is complex, some countries have CBD-related access and benefit sharing legislation in place, and many others do not. Furthermore, access and benefit sharing (ABS) legislation is primarily aimed at the commercial sector with permitting and licensing procedures designated accordingly. Non-commercial activities such as conservation, display, education and research are often grey areas as far as the law is concerned. While carrying out research or conservation activities in collaboration with local partners on the ground is rarely problematic, access to plant material across borders is often subject to complex negotiations or may simply be impossible. Even where material falls under the facilitated access of Annex 1 of the ITPGRFA, it can be difficult to acquire (Bjornstad *et al.*, 2013).

As there is no fast-track, multilateral system for the exchange of non-crop plant material, the most common approach to ABS is adherence to national legislation and acquisition of prior informed consent (PIC) on a country by country basis. In some cases, the negotiation of specific bilateral agreements with partner institutions or governments is necessary, particularly where regulations are still pending or where PIC procedures are cumbersome, and a more programmatic approach is seen by both parties as beneficial. This latter approach is employed by Kew's Millennium Seed Bank and a useful guide to drafting such agreements is given by Cheyne (2003).

Given the complex array of laws and regulations governing access to genetic resources across the world and the expense involved in developing bilateral agreements, it is essential that the botanic garden community develops tools to enable facilitated access to plant material for non-commercial uses such as display, research and conservation.

The International Plant Exchange Network (IPEN) was established for this purpose.[11] The IPEN is a registration system open to botanic gardens that adopt a common policy (code of conduct) regarding access to genetic resources and sharing of the resulting benefits. It was developed by the Verband Botanischer Gärten (an association of gardens in German speaking countries) and was then adopted by the European Botanic Gardens Consortium. The IPEN network facilitates the exchange of plant material between member gardens while respecting the access and benefit sharing regulations of the CBD. It aims to create a climate of confidence between the countries owning the genetic resources and botanic gardens. Gardens that wish to join the network must sign and abide by a code of conduct that sets out gardens' responsibilities for acquisition, maintenance and supply of living plant material and associated benefit sharing. Acquisition or supply of material with extra terms and conditions, or any use for commercial purposes, is not covered by the network and requires the use of appropriate material transfer agreements.

In order to scale up the use of IPEN and to encourage this approach in other networks, BGCI is in the process of developing an advanced search version of its database, PlantSearch,[12] that will ultimately enable botanic gardens to see which garden has what plant material, and which will facilitate access to that material, adhering to the relevant ABS, biosecurity and other statutory requirements of donor and recipient countries. In addition, at the time of writing, *Index Seminum* – a seed exchange system used in many botanic gardens – is being developed into a web-based exchange tool that incorporates information on ABS and plant health regulations and requirements (Havinga et al., 2016).

5.7 The Conservation Opportunities Afforded by *Ex Situ* Collections

All botanic garden collections have value as sources of data or material for research, conservation and use. The utility of seed collections from botanic garden-based seed banks for human innovation and adaptation in agriculture, forestry, horticulture and other sectors is well documented (Smith et al., 2011). Botanic garden seed banks play a critical role in conserving and making seed available for use in areas such as food security (Smith, 2008; Castañeda-Álvarez et al., 2016), combating desertification and water scarcity (e.g. Sacande and Berahmounni, 2016) and ecological restoration (Bozzano et al., 2014).

Other *ex situ* collections, including *in vitro*, cryopreserved and living collections, can also provide materials for these activities. However, they are relatively underutilised for these purposes. Too often the living collections in arboreta and botanic gardens are used simply for display purposes even though they may have a wealth of provenance and other data associated with them that enables detailed study. We tend to forget that botanic gardens in the past were established as living laboratories underpinning economic botany, commerce and land use. The value of well-documented living collections for conservation and use is equally pertinent today.

As one example, mentioned earlier, the International Plant Sentinel Network (IPSN) makes use of living collections in botanic gardens and arboreta to identify new pests and diseases,[13] acting as an early warning system to recognise new and emerging pest and pathogen risks.

Living collections have similar utility as living laboratories in enabling us to understand the abiotic conditions that species will tolerate under domestication, with climate change or as potentially invasive species. The use of bioclimatic envelope modelling based on natural distributions of plant species to predict susceptibility to climate change or, conversely, the potential for a species to become invasive, has limitations particularly where natural distribution is affected by factors other than climate, such as physical barriers and anthropogenic factors (Araujo and Townsend Peterson, 2012). Well-documented, outdoor living collections (as opposed to collections in glasshouses) offer researchers a wealth of information about species' abiotic tolerance and optimal growing conditions (Early and Sax, 2014). Arboreta, in particular, have long been used for this purpose and today that purpose has extra urgency with the threat of climate change, and with assisted migration or translocation increasingly part of the conservationist's armoury. For example, the European Union's REseau INFrastructure de recherche pour le suivi et l'adaptation des FORêts au Changement climatiquE (REINFFORCE) project has established plots in arboreta from Portugal to Scotland to monitor trends in mortality and growth of the most common European tree species under climate change, and taking a long-term perspective.[14]

Living collections in botanic gardens and arboreta are also a useful resource for measuring phenological phenomena in plants, including flowering, fruiting and plant–animal interactions – particularly plant–pollinator relationships. Denver Botanic Gardens, for example, is part of the US National Phenology Network and, as such, uses citizen scientists to monitor phenological trends for 10 species in its garden.[15]

The study of phenology is particularly useful in understanding plant responses to the changing climate, and potential disruption to their life cycles (Harper *et al.*, 2004; Primack and Miller-Rushing, 2009).

Finally, as indicated above, all *ex situ* collections, including seed, *in vitro*, cryopreserved, and living collections, can be useful as sources of data and material for species reintroductions or translocations. Useful information on abiotic tolerance and optimal conditions (e.g. climate and soil) can be obtained from living collections in botanic gardens and arboreta. *In vitro* collections can be used to provide plants for experimental studies on growth and development of endangered taxa, when other material is unavailable or would require using up limited seed stocks or potentially harming plants growing in the wild (Trusty, *et al.*, 2009). *In vitro* collections provide a unique opportunity as a source of plants because, in theory, each tissue culture line can produce plants indefinitely. *Ex situ* collections may also be used a source of plant material for reintroduction or augmentation, particularly when a species or population held in a botanic garden is extinct in the wild. Botanic gardens can use their expertise to propagate plants from their living collections, from stored seed, or from *in vitro* or cryopreserved tissues, to provide plants for restoration. The use of such materials has compared well with wild-sourced materials and can help protect and preserve existing wild populations (Dalrymple *et al.*, 2012). *In vitro* collections are particularly useful for experimental restoration. If a project fails, the reasons can be evaluated and more plants can be produced to test a revised reintroduction protocol, a situation that would deplete valuable stocks if seeds were used (Pence *et al.*, 2008).

Such restoration projects are the ultimate goal of *ex situ* conservation. *Ex situ* collections, by providing a secure backup for species in the event they are threatened with extinction, serve as the leading resource for maintaining species in and restoring them to a wild, evolving habitat. Thus, as the primary stewards of *ex situ* collections of endangered plant taxa worldwide, botanic gardens offer an unmatched resource for conservation of the protection and plant diversity into the future.

Notes

1. See www.bgci.org/plant_search.php.
2. See www.bgci.org/garden_search.php.
3. See http://ncbg.unc.edu/uploads/files/CPCReintroductionPracticeGuidelines.pdf.

4. See www.anbg.gov.au/anpc/publications/translocation.html.
5. See http://data.kew.org/ukgerm/search.
6. See http://data.kew.org/sid/.
7. See www.kew.org/science-conservation/collections/millennium-seed-bank.
8. See www.bgci.org/ourwork/seedupload.
9. See www.bgci.org/plant-conservation/seedawards/.
10. See www.bgci.org/plant-conservation/seedawards/.
11. See www.bgci.org/policy/ipen/.
12. See www.bgci.org/plant_search.php.
13. See www.plantsentinel.org/index/.
14. See http://reinfforce.iefc.net/.
15. See www.botanicgardens.org/science-research/phenology.

References

Araujo, M. B. and Townsend Peterson, A. (2012). Uses and misuses of bioclimatic envelope modelling. *Ecology*, 93(7): 1527–1539.

Ballesteros, D., Estrelles, E., Walters, C. and Ibars, A. M. (2011). Effect of storage temperature on green spore longevity for the ferns *Equisetum ramosissimum* and *Osmunda regalis*. *CryoLetters*, 32: 89–98.

Ballesteros, D., Estrelles, E., Walters, C. and Ibars, A. M. (2012). Effects of temperature and desiccation on *ex situ* conservation of nongreen fern spores. *American Journal of Botany*, 99: 721–729.

BGCI (2013a). BGANZ Collections Assessment. Available online at: http://www.bgci.org/usa/bganz2013/ [accessed March 2017].

BGCI (2013b) Progress Report on Target 8 of the Global Strategy for Plant Conservation in the United States. Available online at: http://www.bgci.org/files/UnitedStates/NACA/NACA_2013_Report.pdf [accessed March 2017].

Bjornstad, A., Tekle, S. and Goransson, M. (2013). 'Facilitated access' to plant genetic resources: does it work? *Genetic Resources Crop Evolution*, 60: 1959–1965, doi 10.1007/s10722-013-0029-6.

Bozzano, M., Jalonen, R., Thomas, E., Boshier, D., Gallo, L., Cavers, S., Bordács, S., Smith, P. and Loo, J., (Eds) (2014). *Genetic Considerations in Ecosystem Restoration Using Native Tree Species. State of the World's Forest Genetic Resources – Thematic Study*. Rome: FAO and Bioversity International.

Brand, J. J. and Diller, K. R. (2004). Application and theory of algal cryopreservation. *Nova Hedwigia*, 79: 175–189.

Castañeda-Álvarez, N. P., Khoury, C. K., Achicanoy, H. A. *et al.* (2016). Global conservation priorities for crop wild relatives. *Nature Plants*, doi: 10.1038/NPLANTS.2016.22.

Cheyne, P. (2003). Access and Benefit-Sharing Agreements: Bridging the Gap between Scientific Partnerships and the Convention on Biological Diversity. In: Smith, R. D., Dickie, J. B., Linington, S. H., Pritchard, H. W. and Probert, R. J. (Eds), *Seed Conservation: Turning Science into Practice*. Richmond, UK: Royal Botanic Gardens, Kew.

Childs, K. H., Tessarolli, L. P. and Day, J. G. (2015). Forty years in liquid nitrogen: an investigation into cryobank management and culture viability. *European Journal of Phycology*, 50: 180.

Christiansen, M. L. (1998). A simple protocol for cryopreservation of moss. *The Bryologist*, 101: 32–35.

Cogălniceanu, G. (2014). Romanian *in vitro* bryophyte collection and its role for conservation. *Acta Horti Botanici Bucurestiensis*, 41: 5–11.

Crane, P. (2013). *Ginkgo: The Tree that Time Forgot*. New Haven and London: Yale University Press.

Dalrymple, S. E., Banks, E., Stewart, G. B. and Pullin, A. S. (2012). A Meta-Analysis of Threatened Plant Reintroductions from Across the Globe. In: Maschinski, J., Haskins, K. E. (Eds), *Plant Reintroduction in a Changing Climate*. Washington DC: Island Press, pp. 31–50.

Day, J. G., Benson, E. E., Harding, K. et al. (2005). Cryopreservation and conservation of microalgae: the development of a Pan-European scientific and biotechnological resource (The COBRA Project). *CryoLetters*, 26: 231–238.

Dereuddre, J., Hassen, M., Blandin, S. and Kaminski, M. (1991). Resistance of alginate-coated somatic embryos of carrot (*Daucus carota* L.) to desiccation and freezing in liquid nitrogen: 2. Thermal analysis. *CryoLetters*, 12: 135–148.

Demidov, A. S. (Ed.) (2012). *Plant Gene Pool of the Red Book of the Russian Federation Conserved in Botanical Gardens and Arboretums*. Moscow: KMK Scientific Press Ltd.

Duckett, J. G., Goode, J. A., and Stead, A. D. (1993). Studies of protonemal morphogenesis in mosses. I. *Ephemerum*. *Journal of Bryology*, 17: 397–498.

Early, R. and Sax, D. (2014). Climatic niche shifts between species' native and naturalized ranges raise concern for ecological forecasts during invasions and climate change. *Global Ecology and Biogeography*, 23: 1356–1365.

ENSCONET (2015). The European Seed Conservation Network. Available online at http://enscobase.maich.gr/ [accessed March 2017].

Fabre, J. and Dereuddre, J. (1990). Encapsulation-dehydration: A new approach to cryopreservation of *Solanum* shoot-tips. *CryoLetters*, 11: 413–426.

FAO (2014). *Genebank Standards for Plant Genetic Resources for Food and Agriculture*. Revised edition. Rome: Commission on Genetic Resources for Food and Agriculture.

Goode, J. A., Stead, A. D. and Duckett, J. G. (1993). Redifferentiation of moss protonemata: An experimental and immunofluorescence study of brood cell formation. *Canadian Journal of Botany*, 71: 1510–1519.

Green, L. A. and Neefus, C. D. (2014). The effects of short- and long-term freezing on *Porphyra umbilicalis* Kützing (Bangiales, Rhodophyta) blade viability. *Journal of Experimental Marine Biology and Ecology*, 461: 499–503.

Guerrant, E. O., Jr., Fiedler, P. L., Havens, K. and Maunder, M. (2004). Revised Genetic Sampling Guidelines for Conservation Collections of Rare and Endangered Plants. In: Guerrant, E.O., Jr, Havens, K. and Maunder, M. (Eds). *Ex situ Plant Conservation: Supporting Species Survival in the Wild*, Washington DC: Island Press, pp. 419–441.

Harper, G. H., Mann, D. G. and Thompson, R. (2004). Phenological monitoring at Royal Botanic Garden Edinburgh. *Sibbaldia*, 2: 35–45.

Havinga, R., Kool, A., Achille, F. et al. (2016). The Index Seminum: Seeds of change for seed exchange. *Taxon*, 65(2): 333–336, doi http://dx.doi.org/10.12705/652.9.

Heesch, S., Day, J. G., Yamagishi, T., Kawai, H., Müller, D. G. and Küpper, F. C. (2012). Cryopreservation of the model alga *Ectocarpus* (Phaeophyceae). *CryoLetters*, 33: 327–336.

Hooke, R. L., Martin-Duque, J. F. and Pedraza, J. (2012). Land transformation by humans: a review. *GSA Today*, 22(12), 4–10, doi: 10.1130/GSAT151A.1.

Keller, E. R. J., Senula, A., Leunufna, S. and Grübe, M. (2006). Slow growth storage and cryopreservation – tools to facilitate germplasm maintenance of vegetatively propagated crops in living plant collections. *International Journal of Refrigeration*, 29: 411–417.

La Farge, C., Williams, K. H. and England, J. H. (2013). Regeneration of little ice age bryophytes emerging from a polar glacier with implications of totipotency in extreme environments. *Proceedings of the National Academy of Sciences*, 110: 9839–9844.

Larkin, P. J. and Scowcroft, W. R. (1981). Somaclonal variation: a novel source of variability from cell cultures for plant improvement. *Theoretical and Applied Genetics*, 60(4): 197–214.

Lee, Y. N. and Nam, K. W. (2016). Cryopreservation of gametophytic thalli of *Ulva prolifera* (Ulvales, Chlorophyta) from Korea. *Journal of Applied Phycology*, 28: 1207–1213.

Li, D.-Z. and Pritchard, H. W. (2009). The science and economics of *ex situ* plant conservation. *Trends in Plant Science*, 14: 614–621.

Li, Y. and Shi, L. (2015). Effect of maturity level and desiccation process on liquid nitrogen storage of green spores of *Osmunda japonica*. *Plant Cell, Tissue and Organ Culture,* 120: 531–538.

Luxoro, C. and Santelice, B. (1989). Additional evidence for ecological differences among isomorphic reproductive phases of *Iridaea laminarioides* (Rhodophyta: Gigartinales). *Journal of Phycology* 25: 206–212.

Maschinski, J., Albrecht, M. A., Monks, L. and Haskins, K. E. (2012). Center for Plant Conservation Best Reintroduction Practice Guidelines, Appendix I. In: Maschinski, J. and Haskins, K.E. (Eds), *Plant Reintroduction in a Changing Climate: Promises and Perils, the Science and Practice of Ecological Restoration*. Washington DC, Island Press.

Matsumoto, T., Sakai, A., Takahashi, C., and Yamada, K. (1995). Cryopreservation of *in vitro*-grown apical meristems of wasabi (*Wasabia japonica*) by encapsulation-vitrification method. *CryoLetters*, 16: 189–196.

Normah, M. N. and Makeen, A. M. (2008). Cryopreservation of Excised Embryos and Embryo Axes. In: Reed, B. M. (Ed.), *Plant Cryopreservation: A Practical Guide*. New York: Springer, pp. 211–233.

Oliver, M. J., Velten, J. and Mishler, B. D. (2005). Desiccation tolerance in bryophytes: A reflection of the primitive strategy for plant survival in dehydrating habitats? *Integrative and Comparative Biology*, 45: 788–799.

Panis, B., Piette, B. and Swennen, R. (2005). Droplet vitrification of apical meristems: A cryopreservation protocol applicable to all *Musaceae*. *Plant Science*, 168: 45–55.

Pence, V. C. (2000a). Survival of chlorophyllous and nonchlorophyllous fern spores through exposure to liquid nitrogen. *American Fern Journal*, 90, 119–126.
Pence, V.C. (2000b). Cryopreservation of *in vitro* grown gametophytes. *American Fern Journal*, 90: 16–23.
Pence, V. C. (2011). Evaluating costs for *in vitro* propagation and preservation of endangered plants. *In Vitro Cellular and Developmental Biology – Plant*, 47: 176–187.
Pence, V. C., Murray, S., Whitham, L., Cloward, D., Barnes, H. and Van Buren, R. (2008). Supplementation of the Autumn Buttercup Population in Utah, USA, Using *In Vitro* Propagated Plants. In: Soorae, P. S. (Ed.), *Global Re-introduction Perspectives*. Abu Dhabi, UAE: IUCN/SSC Re-introduction Specialist Group, pp. 239–243.
Primack, R. B. and Miller-Rushing, A. J. (2009). The role of botanical gardens in climate change research. *New Phytologist*, 182: 303–313.
Pritchard, H. W., Moat, J. F., Ferraz, J. B. S., Marks, T. R., Camargo, J. L. C., Nadarajan, J. and Ferraz, I. D. K. (2014). Innovative approaches to the preservation of forest trees. *Forest Ecology and Management*, 333: 88–98.
Probert, R. J., Daws, M. I. and Hay, F. R. (2009). Ecological correlates of *ex situ* seed longevity: a comparative study on 195 species. *Annals of Botany* 104: 57–69, doi:10.1093/aob/mcp082.
Reed, B. M. (2008). Cryopreservation of Temperate Berry Crops. In: Reed, B. M. (Ed.), *Plant Cryopreservation: A Practical Guide*. New York: Springer, pp. 333–364.
Reed, B.M. and Uchendu, E. (2008) Controlled rate cooling. In: Reed, B. M. (Ed.), *Plant Cryopreservation: A Practical Guide*. New York: Springer, pp. 77–92.
Reed, B. M., Engelmann, F., Dulloo, M. E., and Engels, J. M. M. (2004). *Technical guidelines for the management of field and* in vitro *germplasm collections*. IPGRI Handbooks for Genebanks No. 7, IPGRI (now Bioversity International).
Rowntree, J. K. and Ramsay, M. M. (2005). Ex situ conservation of bryophytes: Progress and Potential of a Pilot Project. *Boletín de la Sociedad Española Briologia*, 26–27: 17–22.
Rowntree, J. K., Pressel, S., Ramsay, M. M., Sabovljevic, A. and Sabovljevic, M. (2011). In vitro conservation of European bryophytes. In Vitro *Cellular and Developmental Biology – Plant*, 47: 55–64.
Sabovljević, M. S., Papp, B., Sabovljević, A. and Segarra-Moragues, J. G. (2012). *In vitro* micropropagation of rare and endangered moss *Entosthodon hungaricus* (Funariaceae). *Bioscience Journal*, 20: 632–640.
Sabovljević, M., Vujičić, M., Pantović, J. and Sabovljević, A. (2014). Bryophyte conservation biology: *In vitro* approach to the *ex situ* conservation of bryophytes from Europe. *Plant Biosystems*, 148: 857–868.
Sacande, M. and Berahmounni, N. (2016). Community participation and ecological criteria for selecting species and restoring natural capital with native species in the Sahel. *Restoration Ecology*, 24(4): 479–488, doi: 10.1111/rec.12337.
Sakai, A. and Nishiyama, Y. (1978). Cryopreservation of winter vegetative buds of hardy fruit trees in liquid nitrogen. *HortScience*, 13: 225–227.
Sakai, A., Kobayashi, S. and Oiyama I. (1990). Cryopreservation of nucellar cells of navel orange (*Citrus sinensis* Osb. Var. *Brasiliensis* Tanaka) by vitrification. *Plant Cell Reports*, 9: 30–33.

Sax, D. F., Early, R. and Bellemare, J. (2013). Niche syndromes, species extinction risks, and management under climate change. *Trends in Ecology and Evolution*, 28(9): 517–52

Schokman, L. (2012). *Plants of the Kampong*. Coconut Grove, FL: National Tropical Botanic Garden.

Sharrock, S., Oldfield, S. and Wilson, O. (2014). *Plant Conservation Report 2014: A Review of Progress in Implementation of the Global Strategy for Plant Conservation 2011–2020*. Richmond, UK: Secretariat of the Convention on Biological Diversity, Montréal, Canada and Botanic Gardens Conservation International, Technical Series No. 81, 56 pp.

Smith, P. P. (2008). *Ex Situ* Conservation of Wild Species: Services Provided by Botanic Gardens. In: Maxted, N., Ford-Lloyd, B. V., Kell, S. P., Iriondo, J. M., Dulloo, M. E. and Turok, J. (Eds), *Crop Wild Relative Conservation and Use*, Wallingford, UK: CABI International. pp.407–412.

Smith, P. P. (2016). Building a Global System for the conservation of all plant diversity: a vision for botanic gardens and Botanic Gardens Conservation International. *Sibbaldia*, 14: 5–13.

Smith, P. P., Dickie, J., Linington, S., Probert, R. and Way, M. (2011). Making the case for plant diversity. *Seed Science Research*, 21, 1–4.

Smith, R. D., Dickie, J. B., Linington, S. H., Pritchard, H. W. and Probert, R. J. (Eds) (2003). *Seed Conservation: Turning Science into Practice*. Richmond, UK: Royal Botanic Gardens, Kew, Richmond, U.K.

The Plant List (2013). A working list of all known plant species, version 1.1. Available online at www.theplantlist.org/ [accessed 5 June 2016].

Trusty, J. L., Miller, I., Pence, V. C., Plair, B. L., Boyd, R. S. and Goertzen, L. R. (2009). *Ex situ* conservation of the federally endangered plant species *Clematis socialis* Kral (Ranunculaceae): a collaborative approach. *Natural Areas Journal*, 29: 376–384.

Tweddle, J. C., Dickie, J. B., Baskin, C. C. and Baskin, J. M. (2003). Ecological aspects of seed desiccation sensitivity. *Journal of Ecology*, 91: 294–304, doi:10.1046/j.1365-2745.2003.00760.x.

Volk, G. M., Bonnart, R., Waddell, J. and Widrlechner, M. P. (2009). Cryopreservation of dormant buds from diverse *Fraxinus* species. *CryoLetters*, 30: 262–267.

Vujičić, M., Sabovljević, A., Sīnzăr-Sekulić, J., Skorić, M. and Sabovljević, M. (2012). *In vitro* development of the rare and endangered moss *Molendoa hornschuchiana* (Hook.) Lindb. Ex Limpr. (Pottiaceae, Bryophyta). *HortScience*, 47: 84–87.

Wallace, S. H. (2015). Development of an Informational Resource to Inform Global Prioritization of Efforts to Conserve Threatened, Exceptional Plant Taxa. Master's Thesis, University of Delaware-Longwood Gardens.

Walters, C., Wheeler, L. M. and Grotenhuis, J. M. (2005). Longevity of seeds stored in a genebank: species characteristics. *Seed Science Research*, 15: 1–20.

Xia, K., Hill L. M., Li, D.-Z. and Walters, C. (2014). Factors affecting stress tolerance in recalcitrant embryonic axes from seeds of four *Quercus* (Fagaceae) species native to the USA or China. *Annals of Botany*, 114: 1747–1759.

6 · The Role of Botanic Gardens and Arboreta in Restoring Plants
From Populations to Ecosystems
KAYRI HAVENS

6.1 Introduction

We live at a time when the world's natural areas are under continual assault; from invasive species, pollution, human use impacts and fragmentation, and climate-related natural disasters like drought, wildfire and flooding. Ecological restoration, the process of assisting the recovery of an ecosystem that has been degraded, damaged or destroyed (SER, 2004), is becoming ever more necessary to ensure that resilient ecological communities, including the biodiversity they support and the ecosystem services they provide, are maintained. The need for increased ecological restoration has been noted in numerous international policy documents including the Global Strategy for Plant Conservation, the Aichi Biodiversity Targets, the Millennium Ecosystem Assessment, and others. Worldwide, governments have already committed to restoring a staggering 3.5 million km^2 by 2030 (UN Climate Summit, 2014) in part to reach the United Nations' target to restore 15 per cent of the world's degraded ecosystems by 2020.

Throughout the over 500-year history of botanic gardens and arboreta (hereafter botanic gardens or gardens), research has been an important part of their activities. Many botanic gardens were originally developed to curate collections in support of taxonomic research or the study of medicinal plants, and most maintain strong programmes in plant systematic and floristic work. In the past few decades, as scientists have recognised the extinction crisis facing us, the missions of many gardens have embraced conservation and restoration roles (Maunder et al., 2004a). As zoos have done for rare animals, gardens have developed *ex situ* conservation and breeding programmes to provide plant material for reintroduction and restoration. With core competencies in plant identification, taxonomy and ecology, as well as vast horticultural knowledge, botanic gardens are well positioned to provide expertise necessary for successful restoration projects. For example, the implementation of a restoration project requires the ecological understanding to know if a degraded site is likely to recover with proper management or whether it requires complete restoration. It requires knowledge of native plant distributions and plant community composition in order to develop appropriate seed mixes based on reference ecosystems. It requires understanding of conditions for seed dormancy break and plant establishment from seed. It requires being able to discern seedlings of native taxa from weeds. Botanic gardens have expertise in all of these areas. Furthermore, as discussed in Chapter 5, many gardens maintain diverse seed banks and living collections that can provide material for use in restoration (Hardwick *et al.*, 2011; Aronson *et al.*, 2014; Miller *et al.*, 2016). The vast majority of seed banks that are focused on wild plant diversity are held at botanic gardens; the largest of these is the Millennium Seed Bank at Royal Botanic Gardens, Kew. In addition, *c.*100,000 species are held in garden living collections globally (Sharrock, 2012).

6.2 Types of Restoration Work Undertaken by Botanic Gardens

6.2.1 Rare Plant Reintroductions

Restoration in the broad sense can occur over many scales from single species population reintroductions to large-scale community and ecosystem restoration projects. Rare plant reintroduction has been an activity championed and carried out by botanic gardens for several decades. With the recognition of plant endangerment and subsequent recovery planning for listed species, many gardens have undertaken single species reintroduction projects. The Center for Plant Conservation (CPC) in the US became a leader in these types of projects and their books on species-level conservation and reintroduction have helped refine and improve projects around the world (Falk and Holsinger, 1991; Falk *et al.*, 1996; Guerrant *et al.*, 2004; Maschinski and Haskins, 2012). The increase in reintroduction attempts and successes over time (Figure 6.1) is testament to the investment made by the CPC and its member gardens in the science of rare plant reintroduction.

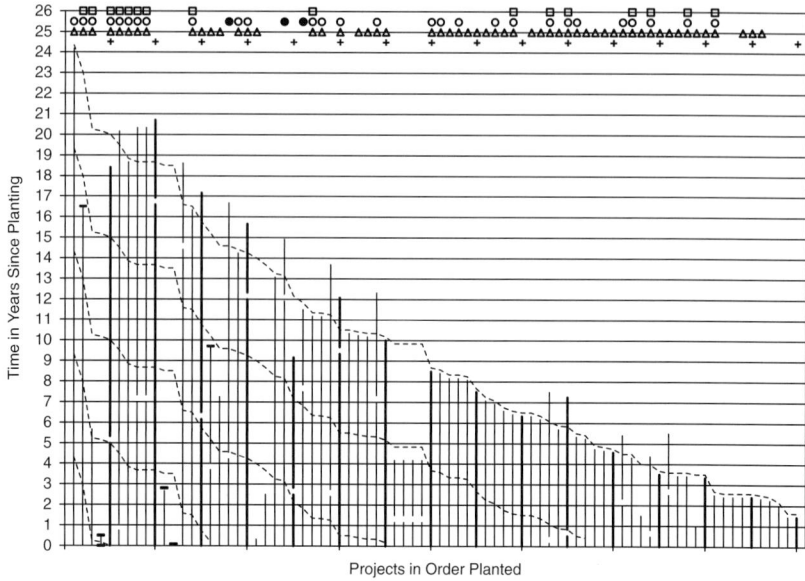

Figure 6.1 Graphical summary of 80 reintroduction projects in order of planting, based on Figure 2.1 in Guerrant (2012). Vertical lines with no marker at the top indicate when a reintroduction was last seen and was alive. Horizontal dashes indicate the last known time a reintroduction known to have failed was still alive. Information about most projects ends just before 2010 and represents our knowledge of these projects at the time of the CPC symposium for which the data were originally gathered. An effort was made in the early months of 2012 to ascertain the fates of projects for which the fate in 2009 was unknown to Guerrant, and the results of these inquiries are placed after short spaces in some vertical lines. Some of these extend beyond 2010, others do not. No attempt was made to follow up on those projects whose fate in 2009 was known. Triangles at top of each column indicate that at least one individual reached reproductive status, a circle indicates that a next generation was produced, and a square if the next generation reached reproductive maturity. See Guerrant (2012) and Guerrant *et al.* (2012) for list of taxa and references. Dashed lines indicate five year iso-chronoclines beginning with 1 January 1990, before which only four projects had been planted, the first in 1986, and ending with the top dashed line indicating 1 January 2010. Reproduced with permission from Guerrant (2013), © Canadian Science Publishing or its licensors.

For example, in the US, over 107 taxa have been reintroduced since 1985, most by botanic gardens (Guerrant, 2012). While the science of reintroduction remains an active field of research, and success is far from certain, reintroduction can be a fundamental part of species recovery. Of the 49 reintroduction projects reported to CPC's International Reintroduction Registry that were initiated between 1985 and 2008,

92 per cent were extant in 2009 (Kennedy *et al.*, 2012). However, many of those projects were quite young and there may have been a reporting bias (people are more likely to enter successful projects into the registry). Nevertheless, we have successes to celebrate, such as the work of the New England Wild Flower Society to restore *Potentilla robbinsiana* which contributed to its removal from the US federal endangered species list (Federal Register, 2002). One of the challenges was developing transplantation protocols for this alpine species after it was grown at a sea-level nursery. Contrary to expectations, transplants of actively growing plants in mid July fared better than those planted immediately from cold storage while dormant (Brumback *et al.*, 2004).

Botanic gardens are having success reintroducing species that are notoriously challenging to grow even under controlled conditions, let alone in natural settings. For example, the Atlanta Botanical Garden (ABG) in Georgia, US, has had success with orchids, known as some of the most difficult plants to work with. The ABG uses its tissue culture laboratory to propagate rare orchids for conservation and restoration. For this, wild-collected orchid seed is germinated in the lab and seedlings are grown in naturalistic environments to prepare them for ultimate reintroduction. One of the rarest orchids in Georgia, US, is the white fringeless orchid (*Platanthera integrilabia*), imperilled due to habitat loss, invasive species and herbivory. Together with a variety of partner organisations, ABG has outplanted more than 100 white fringeless orchids in protected areas (BGCI, 2016a). In the UK, similar efforts undertaken by the Royal Botanic Garden Edinburgh for the fern *Woodsia ilvensis* in Scotland have made it more secure in the wild. After a drastic decline of wild populations, four reintroductions in 1999–2000 established many plants. However, work is ongoing for this species, as there has not been any recruitment observed from reintroduced plants to date (McHaffie, 2006; McHaffie, pers. comm.).

Lack of recruitment is just one reason that even an apparently successful reintroduction does not always guarantee effective long-term conservation for a species. For example, several reintroduction projects have been undertaken for a rare thistle, *Cirsium pitcheri*, by the Morton Arboretum, Chicago Botanic Garden and the United States Geological Service around Lake Michigan in the central United States (see Figure 6.2). While initially quite promising, with demographic analyses indicating that stable, self-sustaining populations had been established, newly emergent threats are casting doubt on the long-term viability of the species. Several species of weevil introduced or distributed as

Figure 6.2 Pitcher's thistle (*Cirsium pitcheri*), a threatened species in the Midwestern USA, is an important pollinator resource. It is threatened in part by several biocontrol agents introduced to control weedy thistles. Photo by Pati Vitt. For the colour version, please refer to the plate section. In some formats this figure will only appear in black and white.

biocontrol agents for weedy thistles have jumped hosts and are negatively impacting this rare, native thistle. Current population projections suggest the species may again be on its way to extinction in the next few decades (Havens *et al.*, 2012). Long-term monitoring, often carried out by gardens, is critical for detecting novel threats like these weevils.

Excellent guidance is provided by Maschinski *et al.* (2012), on whether or not reintroduction is an appropriate conservation measure for a given species, but less attention is given to how to prioritise among the daunting number of taxa that are good candidates. To date, many reintroduction projects have been done on an ad hoc basis, with species selection based on interest and availability. Rare plant reintroductions have been much more common in herbaceous species than trees, despite the importance of trees in providing habitat and other ecosystem services, as well as benefiting people through the provision of food, fuel, timber and more. Only 16 per cent of the reintroduction case studies analysed by Guerrant *et al.* (2012) were woody species, with most of those being shrubs; disproportionately fewer than the estimated 45 per cent of vascular plants that are woody (FitzJohn *et al.*, 2014). Trees, because of their frequent exploitation for timber harvest, are good candidates for 'enrichment planting' using planting stock derived from a botanic garden or arboretum collection (Oldfield and Newton, 2012). Prioritisation of rare plant reintroduction could be improved by focusing on some of the same criteria that guide seed banking decisions; for example, the '5 Es': endangerment, endemism, ecological importance, economic importance and emblematic/educational importance (Maunder *et al.*, 2004b). Partnerships with *in situ* conservation agencies and land stewards can also help prioritise and implement rare plant reintroduction more effectively and sustainably.

6.2.2 Community and Landscape Restoration

Scaling up from single species projects, botanic gardens have also been important players in research on, and the practice of, plant community restoration. For example, the University of Wisconsin Arboretum is the site of Aldo Leopold and others' pioneering work on prairie restoration. The Arboretum's Curtis Prairie is widely cited as the oldest restored prairie dating back to over 80 years (see Figure 6.3). The site was formerly a horse pasture, purchased by the University of Wisconsin for research in 1932/1933 (Court, 2012). Fassett and Thomson began the project in 1935, experimenting with various means of establishing native plants from seeding to transplanting sod. Leopold and others worked with the Civilian Conservation Corps to add more species and reduce exotic species. The use of prescribed fire was initiated experimentally in the late 1930s and continues to be used as a management tool. John Curtis, for whom the prairie is named, was instrumental in setting up long-term monitoring of the restoration which continues to the present day. Invasive species remain an ever-present threat and because of this, the restoration requires, and will continue to need, ongoing management (Wegener *et al.*, 2008). Nevertheless, the Curtis Prairie has provided a window of what the tallgrass prairie once looked like for millions of arboretum visitors over the past several decades and this educational role is a critical part of its mission.

140 · Kayri Havens

Figure 6.3 The University of Wisconsin Arboretum's Curtis Prairie in Madison, WI, USA. Photo by Molly Fifield-Murray. For the colour version, please refer to the plate section. In some formats this figure will only appear in black and white.

Around the world, gardens are sites where land management and horticulture have merged to create small to moderate scale restoration projects that have informed best practices and demonstrated the benefits of native plant use and community restoration for a variety of habitat types. For instance, Royal Botanical Gardens Hamilton and Burlington, in Ontario, Canada, owns and manages approximately 2,400 acres of natural areas that have been under restoration for several decades. Habitats in this reserve include wetlands, forests, savannahs and prairies, but the majority of the site is comprised of Cootes Paradise Marsh, a dedicated bird sanctuary. Restoration has focused on improving water quality, reducing invasive species (including introduced carp, *Cyprinus carpio*) and replanting genetically-appropriate native aquatic plants (BGCI, 2016a).

In Mexico, the Francisco Javier Clavijero Botanic Garden at the Mexican Ecology Institute in Veracruz, Mexico, is working on ecosystem restoration of the cloud forest in Mexico which is one of the world's most threatened habitats. The main goals of this restoration project are to decrease deforestation and restore damaged areas. Local botanists and horticulturists are collaborating to develop practices for germinating and

Role of Botanic Gardens & Arboreta in Restoring Plants · 141

Figure 6.4 Kadoorie Farm Botanic Garden (KFBG) is using experimental plots to evaluate the best horticultural practices for re-establishing forest on Tai Mo Shan, Hong Kong's highest mountain. This grassland site was cleared of forest centuries ago and is covered with abandoned terraces formerly used for tea cultivation. Photo by Stephen Blackmore. For the colour version, please refer to the plate section. In some formats this figure will only appear in black and white

growing wild species in nurseries. Propagation programmes on tree species such as *Symplocos coccinea*, *Podocarpus guatemalensis*, *Styrax glabrescens*, as well as the endangered *Magnolia dealbata*, are underway. Over 1,000 seedlings of *M. dealbata* have been produced for reintroduction, more than are currently remaining in the wild (Gutierrez and Vovides, 1997; Cires *et al.*, 2013; BGCI, 2016b).

At Kadoorie Farm and Botanic Garden (KFBG) in Hong Kong, a restoration demonstration project is attempting to recreate the original forest of Hong Kong, or something very close to it (see Figure 6.4). The original subtropical forest of Hong Kong was destroyed centuries ago and did not recover due to continued disturbance by fire and the lack of dispersal agents for many woody species. This forest would have had over 150 woody species per hectare but has been replaced by fire-maintained grasslands or species-poor secondary forests and scrublands. The KFBG restoration team has used an experimental approach to test different silvicultural treatments, such as the use of different types of fertilisers, biochar, tree guards, mulch, etc. to establish pioneer and climax trees under harsh environmental conditions. To re-establish a stable community so that human intervention can eventually be phased out, late-successional trees were also planted, including thick-leaved oak (*Cyclobalanopsis edithiae*), bamboo oak (*Cyclobalanopsis neglecta*) and tanoak (*Lithocarpus glaber*), as well as two magnolias, Ford's manglietia (*Manglietia fordiana*) and Maud's michelia (*Michelia maudiae*). Over 50,000 saplings have been planted with the hope of restoring a functioning native forest, and all the biodiversity it supports, in Hong Kong (Blackmore, 2014). In addition, this restoration is providing an interesting test of ecological questions related to differences in natural versus artificial succession and the impacts caused by both types of succession on biodiversity and community assembly. Lessons learned will be used for projects elsewhere in the South China Region and will be relevant to human-altered forests around the world.

Beyond their own walls, gardens are also contributing their expertise to larger scale ecological restoration projects. From rehabilitating mine sites and oil fields to restoring forests and grasslands destroyed by wildfire, invasive species, and more, gardens are translating lessons learned on their own sites to landscape level projects. For example, the South China Botanic Garden has a forest restoration project focusing on integrating native tree species such as *Machilus chekiangensis, Phoebe bournei* and *Aquilaria sinensis* in large-scale pine, eucalyptus and acacia plantations of Heshan County, Guangdong, China. Augmentation of the understory vegetation with native species is also underway, including the addition of some medicinal species (e.g. *Millettia speciosa, Gardenia jasminoides, Polygonum cuspidatum, Flemingia stricta*) which may provide both ecological and economic benefits (ERA, 2016a).

The Missouri Botanical Garden (MBG) has been working on a variety of conservation projects in Madagascar for many years. One of their projects focuses on the restoration of a forest plot called Ankofobe. Using herbarium data, MBG scientists have documented the species composition of the site prior to degradation and are using this as a reference to restore the forest and ultimately the ecosystem services it provides, such as clean water for downstream agricultural lands and habitat for several lemur species (Miller *et al.*, 2016). Similarly, King's Park and Botanic Garden (KPBG) in Perth, Australia has done pioneering work on mine site restoration in Western Australia and other arid lands around the world (see Case Study 6.1).

Figure 2.1 Schematic illustrating the relationship between biological complexity and the number of genetic loci required. From left to right, Amazon rainforest Alberto E. Rovi; *Herbertus* David Genney; *Inga edulis* Paul Latham; switchgrass S. E. Wilco, *Mimulus guttatus* A. Twyford.

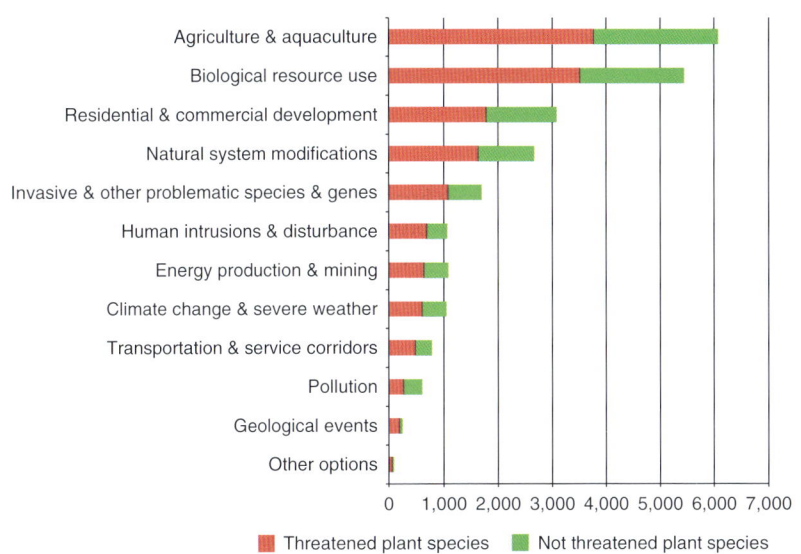

Figure 3.1 The number of plant species (threatened and not threatened) on the IUCN Red List for each major threat (Source: IUCN Red List 2016).

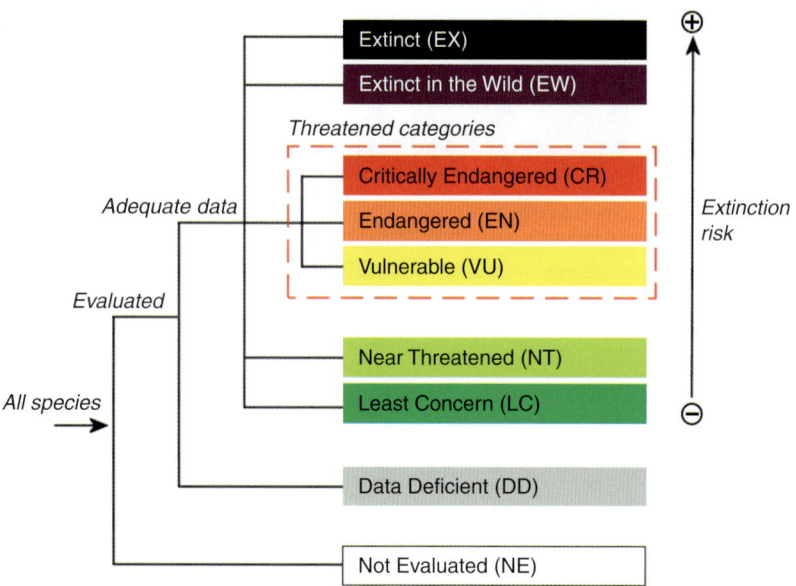

Figure 3.2 IUCN Red List Categories of Threat. Reused with permission from IUCN.

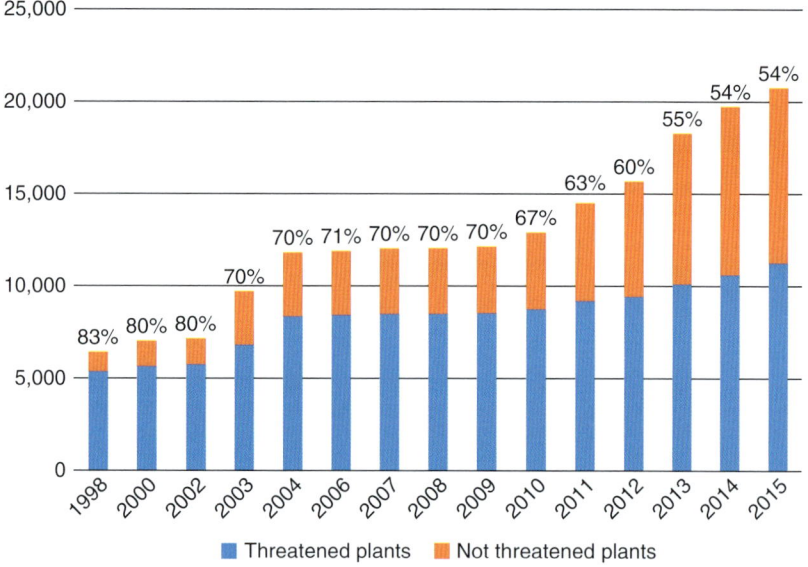

Figure 3.3 The number of plant species (threatened and not threatened) on the IUCN Red List over time, with the percentage of threatened plant species.

Figure 6.2 Pitcher's thistle (*Cirsium pitcheri*), a threatened species in the Midwestern USA, is an important pollinator resource. It is threatened in part by several biocontrol agents introduced to control weedy thistles. Photo by Pati Vitt.

Figure 6.3 The University of Wisconsin Arboretum's Curtis Prairie in Madison, WI, USA. Photo by Molly Fifield-Murray.

Figure 6.4 Kadoorie Farm Botanic Garden (KFBG) is using experimental plots to evaluate the best horticultural practices for re-establishing forest on Tai Mo Shan, Hong Kong's highest mountain. This grassland site was cleared of forest centuries ago and is covered with abandoned terraces formerly used for tea cultivation. Photo by Stephen Blackmore.

Figure 6.5 An intact *Banksia* woodland (a) and sand pit mine restored with *Banksia* woodland species (b) at Gnangara, Western Australia. The last photo (c) shows the establishing seed orchard. Restoration work was done by Kings Park Science, Botanic Gardens and Parks Authority. Photos by Ben Miller.

Figure 6.6 Chicago Botanic Garden Seeds of Success interns, Samantha Primer and Sarah Chambliss collect seed of soaptree yucca (*Yucca elata*). Photo by Mike Howard, Bureau of Land Management, New Mexico, USA.

Figure 6.7 Local herders who work with the Royal Botanic Garden Jordan's Community-Based Rangeland Rehabilitation Programme after a workshop on improving livelihoods and sustainable ecosystem management. Photo by Khalid Al Khalidi.

Figure 8.1 Windy City Harvest Youth Farm students selling at a local farmers' market.

Figure 8.2 Corps participant harvests lettuce on the McCormick Place rooftop in Chicago, IL.

Figure 8.3 Conservation and Land Management interns carry out a quadrat study in the prairie.

Figure 8.4 College First students collect macroinvertebrates in the lagoons at the Chicago Botanic Garden.

Case Study 6.1 *Restoration Research at Kings Park Science in Western Australia*

By Ben Miller, Botanic Gardens and Parks Authority

Kings Park Science is part of the Western Australia (WA) state government's Botanic Gardens and Parks Authority (BGPA). The BGPA has engaged with ecological restoration in two broad fronts: the restoration of degraded areas of native vegetation within the lands it manages (Kings Park and Bold Park), and research both to assist that work and to improve restoration capacity generally in diverse ecosystems across WA and in similar environments around the world. Funding for this research has three main sources: BGPA's own strategic research investment, competitive grant programmes such as the Australian Research Council (ARC) and industry partnerships.

The development of Kings Park Science involved research to cultivate native plants for the WA State Botanic Garden (established 1966) and for the conservation of threatened WA plant species. This research focus complemented and was enhanced by two major initiatives to restore degraded lands managed by BGPA, the Mt Eliza Scarp and Bold Park, from the late 1990s. In recent years, restoration research has chiefly occurred in three regions of WA: the Mediterranean-climate Swan Coastal Plain; the semi-arid Midwest and goldfields; and the subtropical, semi-arid Pilbara. International recognition of this research has led to the extension of restoration research projects overseas, most notably in the Middle East.

One of Kings Park's longest-standing and most productive research partnerships has been with a silica and construction sand mine (Rocla Quarry Products, now Hanson) located at Gnangara, 40 km NNE of Perth on the Swan Coastal Plain. Here, the mining process involves progressive removal of existing vegetation – which was initially native *Banksia* woodlands but has recently moved into aged pine plantations – extraction of resources, and rehabilitation of the sandy pit floors. The presence of high-grade silica ore signifies the almost complete lack of carbon and other minerals in the soil profile and substrate. These *Banksia* woodland soils are the most nutrient-poor mature soils in the world. As the soils and the ecosystem are similar to those managed by BGPA within Kings Park, findings from the

Gnangara site have proven useful for management and restoration of BGPA lands, and vice versa.

Research at Gnangara has resulted in recognition in environmental awards and is held up as a case study in successful post-mining rehabilitation. The scientific value of the historic rehabilitation research and chronosequence is currently being considered as a factor in realignment of a major freeway extension. The demonstrated capacity in the restoration of *Banksia* woodlands at Gnangara has also directly led to policy outcomes, with new standards being set for post-mining restoration of biodiverse ecosystems across the state. Research at the site started with programmes to optimise use of native topsoil, soil seed banks and development of soil profiles for native species return, including treatments such as deep ripping and smoke application (Rokich *et al.*, 2000; Rokich *et al.*, 2001; Rokich *et al.*, 2002). Topsoil seed banks are a critical resource in this system, but supplementary seed broadcasting and greenstock planting have proven necessary to return target biodiversity (Griffiths and Stevens, 2013; Rokich *et al.*, 2002).

Recognising this need, Kings Park researchers began work on understanding seed dormancy and germination processes, which are diverse and complex in the Western Australian flora. Enabling restoration at the landscape scale involves both the study of processes to improve germination and seedling emergence and mechanised seed delivery (Turner *et al.*, 2006; Merritt and Dixon, 2011; Merritt *et al.*, 2014). In the Gnangara area, where pine clearing massively outstrips mining impacts, estimates of the need for rehabilitation of *Banksia* woodlands confirm the need to optimise rehabilitation for scale (Turner *et al.*, 2006). Knowing that 10–15 kg of native seed/ha is required for low to moderate reinstatement of diversity implies a likely demand for 170–255 tonnes of seed required (currently collected at 1 tonne/year) with a value of US$260 million for seed (Stevens *et al.*, 2016).

Another research theme at Kings Park for restoration of *Banksia* woodlands, that incorporates Gnangara, Kings Park and Bold Park sites, employs molecular genetic tools to investigate consequences of seed sourcing and population reconstruction in restoration. This includes understanding the benefits of local seed sources for restoration, the delineation of local provenance, the genetic fitness of source populations, the consequences of mixing provenances, the management of genetic diversity within restored populations, and the ecological genetic assessment of restoration success through the delivery of pollinator services for reproductive functionality and genetic connectivity with adjacent natural remnants (Krauss and Koch, 2004; Krauss et al., 2013; Krauss 2016). To meet on-site seed demand, a seed orchard has been established at Gnangara for five tree species with guidance from these provenance studies.

As establishment of restored sites progressed at Gnangara, new problems and research opportunities arose. Despite early work demonstrating the importance of deep ripping of pit floors, soil compaction remained an ongoing challenge for *Banksia* woodland restoration at Gnangara (see Figure 6.5). Application of experimental treatments and ecophysiological tools have helped to understand this early-establishment problem (Benigno et al.,

Figure 6.5 An intact *Banksia* woodland (a) and sand pit mine restored with *Banksia* woodland species (b) at Gnangara, Western Australia. The last photo (c) shows the establishing seed orchard. Restoration work was done by Kings Park Science, Botanic Gardens and Parks Authority. Photos by Ben Miller. For the colour version, please refer to the plate section. In some formats this figure will only appear in black and white.

2012; Benigno *et al.*, 2013; Benigno *et al.*, 2014). Remote sensing and community ecology metrics have been applied to the restoration chronosequence in order to attempt to untangle the diverse factors that lead to varying rehabilitation outcomes, including site treatments and climate variation (Mounsey, 2014). The maturation of plant populations has also enabled studies of how pollinator communities re-establish and how restored sites add to landscape connectivity (Ritchie and Krauss, 2012; Frick *et al.*, 2014). After more than 20 years of restoration and research at Gnangara, Kings Park and the industry partner are publishing a rehabilitation book for the ecosystem: *Banksia Woodlands: A Guide to their Restoration on the Swan Coastal Plain* (Stevens *et al.*, 2016).

With the research capacity and experience developed from this and other study sites, Kings Park has been invited to assist with development of restoration research programmes internationally. Many of these have been in arid lands in the Middle East and North Africa. The most significant of these is in the Arriyadh Province of central Saudi Arabia. With the Arriyadh Development Authority, Kings Park has established seed banking and seed collecting facilities, resolved germination, enhanced plant propagation capacity and processes and installed and monitored a 6 hectare restoration experiment. The experiment aims to identify optimal irrigation and plant treatment regimes for establishment of four Acacia tree species in an arid regional park.

6.2.3 Restoration Research

Gardens are also instrumental in conducting research that underpins successful restoration. Encompassing diverse disciplines such as seed biology, soil ecology, pollination biology, community ecology and population genetics, botanic garden research frequently focuses on the application of their research to real world conservation and restoration challenges. For example, the Chicago Botanic Garden (CBG) has worked with the Colorado Plateau Native Plant Program on research that aims to improve restoration outcomes in the western United States where much of the landscape has been degraded by invasive species, heavy grazing, drought and altered fire regimes. The programme identifies native species and ecotypes that compete well with invasive species and tolerate disturbance. These early seral species (also known as 'native winners'; Barak *et al.*, 2015) may prove superior to more conservative species for restoration of highly disturbed sites. The CBG scientists are conducting greenhouse and field trials to better understand local adaptation to inform seed sourcing decisions, as well as seed ecology to assist with restoration application decisions.

Another focus of research is the impact of diversity on restorations (see Box 6.1). Many researchers have found positive effects of increased

---

Box 6.1 *Diversity in All its Forms*

We know 'diversity' is important in plant communities and we strive to create diverse restorations, but what do we mean by diverse? There are many measures of diversity that may be important for restorations. They include:

At the species level:
Genetic diversity – Genetic diversity is needed for populations to respond to environmental and evolutionary challenges. Related to restoration, genetic diversity is important in both initial population establishment and long-term success, but attention must be paid to establishing appropriate diversity (i.e. take local adaptation into consideration) and not just aim for maximum diversity (McKay *et al.*, 2005; Kramer and Havens, 2009).
Age/stage diversity – Perennial plant populations are typically stage (i.e. seedling, juvenile, adult) and age structured. Restoring populations with temporal variation in population structure can be important to buffer against environmental stressors such as drought and herbivory (McEachern *et al.*, 1994).

At the community level:
Species diversity – Increased species richness (the number of species in a community) has been correlated with greater levels of ecosystem functioning in many systems (Schwartz *et al.*, 2000; Naeem *et al.*, 2009; Gamfeldt *et al.*, 2013), although this relationship is not universal and has been examined in only a limited number of ecosystem types.
Functional/functional trait diversity – Functional diversity is frequently defined as 'the value and the range of those species and organismal traits that influence ecosystem functioning' and therefore by definition, it is the component of biodiversity that is important for ecosystem stability and function (Tilman, 2001). It is a measure of the relative magnitude of species differences and similarities (Cadotte *et al.*, 2011). Functional

diversity is often correlated with species richness, but not necessarily in a linear fashion (Diaz and Cabido, 2001; Cadotte *et al.*, 2011; Clark *et al.*, 2012).

Phylogenetic diversity – Phylogenetic diversity may be important for ecosystem function, in part because phylogenetically diverse communities may represent more functional traits. High phylo-diversity has also been correlated with communities that are more stable, productive and resistant to invasion and herbivory (Hipp *et al.*, 2015; Navarro-Cano *et al.*, 2016).

Phenological diversity – High phenological diversity, i.e. having plants that provide floral and fruit resources across the entire growing season, may be particularly important for pollinators and other wildlife (Armstrong *et al.*, 2016; Havens and Vitt, 2016).

species diversity on productivity, resistance to invasion, resilience after disturbance and diversity of higher trophic levels such as insects (Tilman and Downing, 1994; Hooper *et al.*, 2005). An intriguing area of research uses the same principles to examine the effect of phylogenetic diversity on improving restorations. As opposed to simply counting the number of species, phylogenetic diversity takes into account the evolutionary history of a plant community, with high phylogenetic diversity in sites with many distantly related species. Researchers are increasingly considering phylogenetic diversity in restoration plans (Hipp *et al.*, 2015). For example, a group of researchers from the Chicago Botanic Garden and the Morton Arboretum is investigating if incorporating phylogenetic diversity into tallgrass prairie restorations improves outcomes. By planting experimental plots with varying amounts of phylogenetic diversity, researchers will determine if phylogenetic diversity impacts productivity, species persistence and invasion resistance. The results will ultimately inform seeds mixes used for restoration and could be applicable in ecosystems beyond tallgrass prairie.

Getting seeds to germinate and establish has been problematic in restoration projects in many habitats and geographic locations. Kings Park and Botanic Garden has been a global leader on research related to seed use. They have conducted research on the role of smoke in breaking seed dormancy in plants of fire-prone habitats. Their work was key to isolating a group of compounds from plant-derived smoke, karrikins, which have improved seed germination in plants from a diversity of ecosystems around the world (Flematti *et al.*, 2004; Chiwocha *et al.*, 2009; Jefferson *et al.*, 2014). They have further studied the storability of seeds that have had dormancy broken ('restoration ready' seed) and found that dormancy break does not change viability and makes the seed easier to use (Turner *et al.*, 2013). In addition, KPBG is developing seed pelleting technology that packages together dormancy-breaking compounds with moisture retention materials to help ensure seed has the best chance possible of germinating and establishing in harsh environments (Erickson *et al.*, 2016). Similar approaches are well underway in the western US to improve sagebrush steppe restoration (Madsen *et al.*, 2016).

In addition to research related to improving the practice of restoration, many gardens are undertaking research related to demonstrating and valuing the ecosystem services provided by restored ecosystems. Larkin and colleagues (2014) found that the restoration of a buckthorn (*Rhamnus cathartica*) invaded woodland in the US resulted in higher plant diversity, higher litter mass, reduced soil erosion and other changes associated with improved carbon storage. They also found that these benefits increased with time since restoration (the sites had ongoing management). Some restorationists had been concerned that removal of a substantial amount of woody plant material in buckthorn thickets would decrease carbon sequestration, but it actually led to a net increase in wood biomass, an increase that was also positively correlated with restoration age. Buckthorn trunks and branches are relatively thin and the researchers believe that taking out buckthorn freed large native trees (like *Quercus* spp.) to better reach their growth potential (Larkin *et al.*, 2014). This finding is not unique; a meta-analysis of restoration projects found that they increased the provision of biodiversity (by 44 per cent) and ecosystem services (by 25 per cent), although neither reaching the level of intact ecosystems, at least in the time frame studied. These gains were particularly strong in tropical terrestrial ecosystems (Benayas *et al.*, 2009).

6.3 Overcoming Challenges

There are numerous challenges that will need to be addressed if large-scale restoration is going to succeed. These include:

- having adequate amounts of native seed and seedlings for restoration projects;
- conducting long-term monitoring of reintroduction and restoration projects;

- understanding how to restore communities and ecosystems in a changing climate;
- garnering community support and integrating community needs into restoration projects;
- addressing technical and scientific challenges; and
- building capacity for restoration globally.

Botanic gardens are actively addressing many of these challenges.

6.3.1 The Need for Seed

Genetically appropriate seed of native plant species has emerged as the rate limiting step for most restoration projects globally. Merritt and Dixon (2011) have called for scaling up seed banks, from 'stamp collections' to collections that can meet the need of tonnes of seed for restoration. This call has been echoed by international networks, policy documents and workshops (e.g. Kildisheva et al., 2016). The Society for Ecological Restoration recently formed a new section, the International Network for Seed Based Restoration, dedicated to seed need and seed use. The section brings together scientists, the seed industry, community, non-governmental organisations and governments who are interested in native seed ecology, conservation and use in restoration, as well as the development and use of native plant crops.

In the US, the federal Bureau of Land Management (BLM) is the largest seed purchaser in the western hemisphere, and probably the world. That agency has purchased over 7 million pounds of seed in bad wildfire years and has averaged approximately 3 million pounds of seed over the past two decades (BLM, 2009). Several botanic gardens are partnering with the BLM in a programme called Seeds of Success (SOS) designed to help the nation meet its need for native seed (see Figure 6.6). The programme is collecting wildland seed for restoration, research, conservation and development as native seed crops (Haidet and Olwell, 2015). In 2015, the Plant Conservation Alliance (PCA) released the National Seed Strategy in the US which is a plan designed to get the 'right seed in the right place and the right time' (PCA, 2015). Because restoration needs, particularly in the western US, are largely driven by wildfires which are unpredictable, it has been difficult to have adequate stores of regionally adapted seed available. One of the issues the strategy should address is short-term storage for restoration seed, allowing land managers to buy seed more regularly, leading to a more dependable

Figure 6.6 Chicago Botanic Garden Seeds of Success interns, Samantha Primer and Sarah Chambliss collect seed of soaptree yucca (*Yucca elata*). Photo by Mike Howard, Bureau of Land Management, New Mexico, USA. For the colour version, please refer to the plate section. In some formats this figure will only appear in black and white.

demand for seed growers. If agencies focus on getting regional ecotypes of early seral matrix (aka 'workhorse') species into storage, they will be prepared to stabilise soil immediately after wildfires with appropriate seed that can be augmented in future years. The strategy also calls for provenance trials for these workhorse species to delineate seed zones, or the area within which plant materials can be transferred with little risk of being poorly adapted to their new location. This will allow us to better understand how widely particular ecotypes can be used without problems associated with maladaptation. Seed zone delineation is typically done by growing multiple geographic collections (provenances) of a particular species in multiple common gardens to assess where they do well and where they do not. Botanic gardens can serve as sites for many of the common garden trials such an approach will require.

6.3.2 Long-Term Monitoring

Botanic gardens have also played an important role in long-term monitoring, often involving citizen scientists. Examples include the New

England Plant Conservation Program developed by New England Wild Flower Society, Plants of Concern at Chicago Botanic Garden and Rare Care at the University of Washington Botanic Garden. The South African National Biodiversity Institute (SANBI) manages a programme called the Custodians of Rare and Endangered Wildflowers (CREW), a volunteer programme that monitors South Africa's threatened plants. Volunteers include school children, university students, professionals and retirees. Volunteers may census rare plants, survey for threatened species, adopt sites for conservation management and in some cases, conduct demographic studies. One project in the Northern Cape is using demographic studies to understand how two rare plants, *Bulbinella latifolia* subsp. *doleritica* and *Euryops virgata*, are responding to threats like climate change and livestock grazing (SANBI, 2010).

Botanic gardens can make institutional commitments to such programmes, ensuring long-term monitoring that can go on long beyond the tenure of any individual staff member or student. While much of this monitoring to date has focused on population trends of threatened plant species (both in natural and reintroduced populations), it could be expanded to look at multi-species restoration success as well. Understanding what has worked and what has not is critical for future meta-analyses that will improve reintroduction and restoration practice. Building on the CPC database of reintroduction project outcomes, referred to earlier, and developing a similar clearing house for community restoration outcomes would be useful.

6.3.3 Restoration and Climate Change

Reintroduction and restoration are challenging enough in a static environment and even more so in a changing one. As we grapple with understanding local adaptation, delineating seed transfer zones, and defining appropriate seed use, climate change is adding unwelcome complications. Assisted migration has been suggested as one way to help plants move to more climatically suitable locations if they are unable to migrate on their own and if they appear to lack the capacity to adapt to the new conditions in their home range. In reality, for better or worse, botanic gardens (and the horticultural industry more broadly) have been moving species around for centuries and often have data on their performance in novel climates. Scientists at Missouri Botanical Garden have suggested that purposeful, 'chaperoned' assisted migration experimentation be done using botanic garden collections (Smith *et al.*, 2014). These data will be extremely useful as the conservation and scientific communities develop best practices regarding assisted migration. The concept has been controversial because many see assisted migration as a potential avenue for new invasive species (Riccardi and Simberloff, 2009). Indeed, many species have become invasive when moved outside their historical range, mostly after intercontinental movement, rather than shorter intracontinental moves, although some examples of the latter have occurred. Others argue that assisted migration is too time consuming and expensive to be practical given the overwhelming problem of climate change and the sheer number of species that may need to move (Hunter, 2007). Still others maintain that assisted migration, while not a panacea, could be implemented through ongoing restoration via changing seed sourcing protocols and may result in better adapted plant communities in the future (Vitt *et al.*, 2009; Breed *et al.*, 2013; Havens *et al.*, 2015).

This debate spills over into discussions about appropriate seed sourcing for restoration under climate change. The 'local seed is best' concept has guided ecological restoration since its inception, and for good reason; there are numerous examples of local adaptation and 'home-site advantage' in plants (Leimu and Fischer, 2008; Hereford, 2009). However, many today are beginning to question these assumptions and ask if we should try to include some 'pre-adapted' plant material sourced from regions more similar to what the future climate is likely to be (Sgro' *et al.*, 2011). This predictive provenancing may be beneficial in the future (if one has predicted correctly regarding future climate) providing it does not preclude using material that is adapted to the current climate. At this point, more experimentation is needed to understand the consequences of different provenancing approaches. It is clear, however, that reintroducing genetically diverse, regionally sourced plant materials remains a good management practice, particularly when sourced from environmentally similar locations (Broadhurst *et al.*, 2008). It is also clear that seed zones may need to become more dynamic in the future (Kramer and Havens, 2009). Regardless of seed source, maintaining genetic diversity from seed collection, through a seed increase project, and ultimately into a restoration is challenging. Loss of diversity can occur at every step of the restoration process, from seed collecting and cleaning to germination and growth in a nursery, through to attrition in the restoration itself. Basey and colleagues (2015) provide recommendations on how to avoid genetic loss during the native plant materials development process which should greatly improve the genetic diversity and integrity of restoration materials.

6.3.4 Integrating Community Needs

A successful restoration project must often balance the needs of the plant community with the needs of the human community. Humans are largely responsible for the factors that have led to ecosystem degradation and these factors must be addressed in order for a restoration to thrive. For instance, if the site has been used for grazing or plant harvest in the past, it may not be practical to completely bar human use, but partnerships with the community can lead to use at sustainable levels. For example, the Tooro Botanical Gardens (TBG) in Uganda is beginning a forest restoration project and is working with neighbouring communities to help select tree species for the project. Currently tree planting projects in Africa often focus on non-native species. This project hopes to promote the benefits of native taxa, and one of the criteria for taxon selection is usefulness (for timber, medicine, etc.). The Tooro Botanical Gardens is also involved in promoting sustainable use through education programmes for women and children demonstrating how to harvest some plant parts without damaging the entire plant (Shaw and Oldfield, 2013).

The importance of integrating the community is nicely exemplified by the innovative Community-Based Rangeland Rehabilitation programme at the Royal Botanic Garden of Jordan (see Case Study 6.2).

Case Study 6.2 *Rangeland Restoration at the Royal Botanic Garden of Jordan*

By Mustafa Al Shudiefat and Khalid Al Khalidi, Royal Botanic Garden of Jordan

When the Royal Botanic Garden of Jordan (RBGJ) was first founded, local herders would illegally allow their herds to graze anywhere and everywhere throughout the entire year. They would even cut the fence surrounding the botanic garden and sneak their herds in very early or late at night, so that they could graze inside the site. This was a huge problem, as the RBGJ needed to be able to restore plant cover, conduct vegetation surveys and make biomass estimates without interference from animals. With this in mind, the RBGJ decided to work with local herding families and establish the Community-Based Rangeland Rehabilitation (CBRR) to develop efficient sustainable rangeland management strategies and to improve local livelihoods through sustainable ecosystem management.

Public meetings with livestock owners and key figures in the area were held to discuss the problem, possible solutions, alternative grazing scenarios, and the timing of grazing. These meetings also fostered cooperation and agreement on sustainable land management approaches. The local herders were offered forage in exchange for not grazing on the site, making it possible for RBGJ to conduct vegetation surveys and biomass estimates and determine sustainable stocking rates and grazing scenarios. The CBRR also provided useful and practical training and advice to the community, and began to establish environmentally friendly income generating programmes. In addition, the CBRR taught herders about better health, hygiene and herd management techniques, and facilitated access to veterinary care.

After training, the CBRR allowed the herders to resume grazing on the site at certain times and under specific conditions. This managed grazing arrangement has yielded positive results for both the land and the livestock owners, and can be replicated in small degraded rangeland areas in other parts of the country. Although only five local herding families cooperated fully with the CBRR in the first year, by 2015, some 48 families were participating. The benefits became quickly evident to the early joiners, and by 2009 livestock owners who once grazed the site to bare soil were policing themselves and teaching others.

Managed grazing has allowed natural regrowth of vegetation and a return of wildlife. Limiting the presence of sheep when plants are sprouting has increased the biomass. And soil fertility has improved, through the mixing of organic matter and manure. Biomass increased by 30 per cent from 2008 to 2009, by another 30 per cent from 2009 to 2010, and 10 per cent per year in subsequent years. Some plant species that disappeared from the region years ago have now spontaneously reappeared. The plant species recorded during RBGJ plant surveys increased from 436 in 2006 to 580 in 2012.

By giving attention to animal health, incomes have also improved. For instance, one herder's income rose from $8,200 in 2007 to over $20,000 annually. Another herder with a smaller flock began with a net loss of $496 a year, but now earns over $6,300 per year. The CBRR helped improve the socio-economic status of the herders in the area, decreased poverty, enhanced child education, improved maternal and family health, and led to a more sustainable environment, all of which contribute to the UN Millennium Development Goals. Today, over 400 people benefit directly from the CBRR project, and an additional 1,500 have benefited indirectly. Approximately 200 hectares have been rehabilitated. Given the interest shown by associations, non-governmental organisations and government agencies, the CBRR is planning to share and transfer its expertise to other herding communities.

It is clear that the governance approach and positive socio-economic effects for the local community should be part of any effort to restore ecosystems in forests and rangelands. The CBRR intends to replicate this model in other areas, and develop sustainable grazing protocols that can be used to improve the quality of rangeland habitats and the livelihoods of pastoralists throughout the region. For more information see the project website.[1]

Figure 6.7 Local herders who work with the Royal Botanic Garden of Jordan's Community-Based Rangeland Rehabilitation programme after a workshop on improving livelihoods and sustainable ecosystem management. Photo by Khalid Al Khalidi. For the colour version, please refer to the plate section. In some formats this figure will only appear in black and white.

This project balanced the needs of local farmers for access to grazing lands with the desire to manage a natural area more sustainably, resulting in good outcomes for both the human and plant communities.

6.3.5 Technical and Scientific Challenges

Restoration ecology is a relatively young discipline with numerous areas of active research including fundamental ecological questions related to species interactions, community assembly, landscape ecology and more. However, there are also numerous applied questions related to genetics, seed technologies, invasive species control, understanding best practices in different environments, and defining and measuring success. These applied questions are critical challenges limiting our ability to meet the vast global restoration need. Developing new partnerships and collaborative learning may be one of the best ways to address these challenges quickly. The Ecological Restoration Alliance of Botanic Gardens (ERA), coordinated by BGCI, was formed to assist with this. The ERA is a global consortium of botanic gardens actively engaged in ecological restoration who have agreed to support efforts to scale up the restoration of damaged, degraded and destroyed ecosystems around the world. Currently comprised of 20 member botanic gardens, the ERA is carrying out ecological restoration projects in a diverse range of ecosystems with the overall aim to 'Connect People, Share Knowledge, and Restore Ecosystems.' The ERA's stated goals are to:

1. Work with local partners to set up, maintain and document a series of long term sustainable exemplar restoration projects in diverse biophysical, political, and cultural contexts around the globe that provide training and demonstrate the value of a carefully designed, science-driven approach to sustainable ecological restoration.
2. Improve the quality and volume of science-based ecological restoration practice by deploying scientific and horticultural skills to applied work on the ground.
3. Conduct ecological restoration research, to develop an enhanced knowledge base for restoration and identify and inform best practice.
4. Disseminate research and lessons learnt from projects.
5. Build expertise and restoration capacity through collaborations between botanic gardens, large and small, as well as with partners in local communities, professional societies, academia, industry, government, non-governmental organisations (NGOs) and international bodies.
(ERA, 2016b)

The ERA may also prove to be effective in building recognition about botanic gardens as important partners in both ecological restoration (the practice) and restoration ecology (the science). Gardens have much to offer in scientific, horticultural and land management expertise that will be crucial in improving restoration outcomes. Breaking down barriers between governmental agencies and NGOs, academics and practitioners, professionals and volunteers is essential to move restoration forward quickly and productively and many gardens are already working effectively with these various communities.

6.4 Conclusions

Clearly botanic gardens have been instrumental in both the practice of and research on single species population reintroductions through to plant community and ecosystem restoration. These approaches and scales are not mutually exclusive. In a sense, community restoration is the introduction of multiple species, combined with the management of ecological processes. Lessons learned from single species reintroductions inform restoration, particularly related to population genetics and seed use (Donaldson, 2009). Rare plant species are often incorporated into community restoration projects, particularly once the appropriate framework of 'workhorse' species has been established. Through iterative adaptive management approaches, our work on reintroduction and restoration continues to improve.

In addition to the very concrete role that botanic gardens are playing in conducting ecological restoration, they also have had a critical role in demonstrating the benefits of restoration to public audiences. The practice of restoration often has some stages that appear destructive to those who are not familiar with the process. Invasive species removal is often the first step and can leave the landscape looking barren. In the Chicago region, there was a well-documented backlash against ecological restoration in the late 1990s, led by those who objected to the removal of brush (largely the invasive shrub, common buckthorn, *Rhamnus cathartica*) from public land. Ecologists, with professional and volunteer stewards, were attempting to restore a more open oak savannah community typical of what occurred in the region pre-European settlement from highly degraded, invaded woodlands. However, to the uninitiated, the process looked like the forests were being removed from the forest preserves (Woodworth, 2013). As a response to the backlash, Chicago Wilderness, a coalition of environmental organisations and agencies, recognised the need to designate 'restoration demonstration sites' to show the public what restoration sites look like further along in the process. Not surprisingly, due to its high visitation and high quality,

Chicago Botanic Garden's McDonald Woods was designated as one of the demonstration sites.

As with many environmental endeavours, funding and capacity remain limited for ecological restoration. Despite the clear need for restoring more resilient landscapes, the public and political will to invest in conservation and restoration is limited. Until we change perceptions through education and advocacy about the critical importance of restoration, we cannot expect funding levels to increase. We need to better demonstrate the importance of diverse native landscapes for wildlife habitat, for ecosystem services like carbon sequestration, flood control, and erosion prevention, for recreational opportunities, for sustainable use sites, and for much more. And with collectively over 200 million visitors per year, botanic gardens are an ideal place to do this.

Note

1. www.royalbotanicgarden.org/page/community-based-rangeland-rehabilitation.

References

Armstrong, J. B., Takimoto, G. T., Schindler, D. E., Hayes, M. M. and Kauffman, M. J. (2016). Resource waves: phenological diversity enhances foraging opportunities for mobile consumers. *Ecology*, 97(5):1099–1112, doi: 10.1890/15–0554.

Aronson, J., On Behalf of The ERA of Botanic Gardens. (2014). The ecological restoration alliance of botanic gardens: a new initiative takes root. *Restoration Ecology*, 22: 713–715.

Barak, R. S., Fant, J. B., Kramer, A. T. and Skogen, K. A. (2015). Assessing the value of potential 'native winners' for restoration of cheatgrass-invaded habitat. *Western North American Naturalist*, 75: 58–69.

Basey, A. C., Fant, J. B. and Kramer, A. T. (2015). Producing native plant materials for restoration: 10 rules to collect and maintain genetic diversity. *Native Plants Journal*, 16: 37–53.

Benayas, J. M. R., Newton, A. C., Diaz, A. and Bullock, J. M. (2009). Enhancement of biodiversity and ecosystem services by ecological restoration: a meta-analysis. *Science*, 325: 1121–1124.

Benigno, S. M., Cawthray, G. R., Dixon, K. W. and Stevens, J. C. (2012). Soil physical strength rather than excess ethylene reduces root elongation of Eucalyptus seedlings in mechanically impeded sandy soils. *Plant Growth Regulation*, 68: 261–270.

Benigno, S. M., Dixon, K. W. and Stevens, J. C. (2013). Increasing soil water retention with native-sourced mulch improves seedling establishment in post-mine Mediterranean sandy soils. *Restoration Ecology*, 21: 617–626.

Benigno, S. M., Dixon, K. W. and Stevens, J. C. (2014). Seedling mortality during biphasic drought in sandy Mediterranean soils. *Functional Plant Biology*, 41: 1239–1248.

BGCI (2016a). *North American Botanic Garden Strategy for Plant Conservation, 2016–2020*. Illinois, US: Botanic Gardens Conservation International.

BGCI (2016b). Botanic Garden Conservation International website, available online at: http://www.bgci.org/worldwide/article/120/ [accessed 25 April 2016].

Blackmore, S. (2014). Seeds of hope on the mountainside. *Resurgence and Ecologist*, 287: 38–39.

BLM (2009). *Native Plant Materials Development Program: Progress Report for FY2001–2007*. Washington DC: US Department of Interior, Bureau of Land Management, 42 pp.

Breed, M. F., Stead, M. G., Ottewell, K. M., Gardner, M. G. and Lowe, A. J. (2013). Which provenance and where? Seed sourcing strategies for revegetation in a changing environment. *Conservation Genetics*, 14: 1–10.

Broadhurst, L. M., Lowe, A., Coates, D. J., Cunningham, S. A., McDonald, M., Vesk, P. A. and Yates, C. (2008). Seed supply for broadscale restoration: Maximizing evolutionary potential. *Evolutionary Applications*, 1: 587–597.

Brumback, W. E., Weihrauch, D. M. and Kimball, K. D. (2004). Propagation and transplanting of an endangered alpine species, Robbins' cinquefoil, *Potentilla robbinsiana* (Rosaceae). *Native Plant Journal*, 5: 91–97.

Cadotte, M. W., Carscadden, K. and Mirotchnick, N. (2011). Beyond species: functional diversity and the maintenance of ecological processes and services. *Journal of Applied Ecology*, 48: 1079–1087.

Chiwocha, S. D. S., Dixon, K. W., Flematti, G. R., Ghisalberti, E. L., Merritt, D. J., Nelson, D. C., Riseborough, J. M., Smith, S. M., Stevens, J. C. (2009). Karrikins: a new family of plant growth regulators in smoke. *Plant Science*, 177: 252–256.

Cires, E., De Smet, Y., Cuesta, C., Goetghebeur, P., Sharrock, S., Gibbs, D., Oldfield, S., Kramer, A. and Samain, M. S. (2013). Gap analyses to support *ex situ* conservation of genetic diversity in Magnolia, a flagship group. *Biodiversity and Conservation*, 22: 567–590.

Clark C. M., Flynn, D. F. B., Butterfield, B. J. and Reich, P. B. (2012). Testing the link between functional diversity and ecosystem functioning in a Minnesota grassland experiment. *PLoS ONE*, 7:e52821.

Court, F. E. (2012). *Pioneers of Ecological Restoration*. Madison, WI. University of Wisconsin Press, 314 pp.

Diaz, S. and Cabido, M. (2001). Vive la difference: plant functional diversity matters to ecosystem processes. *Trends in Ecology and Evolution*, 16: 646–655.

Donaldson, J. S. (2009). Botanic gardens science for conservation and global change. *Trends in Plant Science* 14: 608–613.

ERA (2016a). Ecological Restoration Alliance of Botanic Gardens website. Available online at: http://www.erabg.org/project/35/ [accessed 27April 2016].

ERA (2016b). Ecological Restoration Alliance of Botanic Gardens website. Available online at: http://erabg.org/introduction/ [accessed 27April 2016].

Erickson, T., Barrett, R., Merritt, D. and Dixon, K. (2016). *Pilbara Seed Atlas and Field Guide: Plant Restoration in Australia's Arid Northwest.* Clayton, Victoria: CSIRO Publishing, 312 pp.

Falk, D. A. and Holsinger, K. E. (Eds) (1991). *Genetics and Conservation of Rare Plants.* New York: Oxford University Press, 283 pp.

Falk, D. A., Millar, C. I. and Olwell, M. (Eds). (1996). *Restoring Diversity: Strategies for Reintroduction of Endangered Plants.* Washington, DC: Island Press, 505 pp.

Federal Register (2002). Endangered and threatened wildlife and plants; removal of *Potentilla robbinsiana* (Robbins' cinquefoil) from the federal list of endangered and threatened plants. Final Rule. Vol. 67, No. 166.

FitzJohn, R. G., Pennell, M. W., Zanne, A. E., Stevens, P. F., Tank, D. C. and Cornwell, W. K. (2014). How much of the world is woody? *Journal of Ecology*, 102: 1266–1272.

Flematti, G. R., Ghisalberti, E. L., Dixon, K. W. and Trengove, R. D. (2004). A compound from smoke that promotes seed germination. *Science*, 305: 977.

Frick, K. M., Ritchie, A. L. and Krauss, S. L. (2014). Field of dreams: Restitution of pollinator services in restored bird-pollinated plant populations. *Restoration Ecology*, 22: 832–840.

Gamfeldt, L., Snäll, T. Bagchi, R. *et al.* (2013). Higher levels of multiple ecosystem services are found in forests with more tree species. *Nature Communications*, 4: 1340, doi: 10.1038/ncomms2328.

Griffiths, E. and Stevens, J. C. (2013). Managing nutrient regimes improves seedling root-growth potential of framework banksia-woodland species. *Australian Journal of Botany*, 61: 600–610.

Guerrant, E. O., Jr. (2012). Characterizing Two Decades of Rare Plant Reintroductions. In: Maschinski, J. and Haskins, K. E. (Eds), *Plant Reintroduction in a Changing Climate: Promises and Perils.* Washington, DC: Island Press, pp. 9–29.

Guerrant, E.O., Jr. (2013). The value and propriety of reintroduction as a conservation tool for rare plants. *Botany*, 91: v–x.

Guerrant, E. O., Havens, K. and Maunder, M. (Eds) (2004). *Ex Situ Plant Conservation: Supporting Species Survival in the Wild.* Washington, DC: Island Press, 504 pp.

Guerrant, E. O., Jr., Albrecht, M. A. and Dalrymple, S. (2012). Studies used for meta-analyses. Appendix 2. In: Maschinski, J. and Haskins, K. E. (Eds), *Plant Reintroduction in a Changing Climate: Promises and Perils.* Washington, DC: Island Press, pp. 307–317.

Gutierrez, L., and Vovides, A. P. (1997). An *in situ* study of *Magnolia dealbata* Zucc. in Veracruz State: an endangered endemic tree of Mexico. *Biodiversity and Conservation*, 6: 89–97

Haidet, M. and Olwell, P. (2015). Seeds of success: a national seed banking program working to achieve long-term conservation goals. *Natural Areas Journal*, 35: 165–173.

Hardwick, K. A., Fiedler, P., Lee, L. C., Pavlik, B., Hobbs, R. *et al.* (2011). The role of botanic gardens in the science and practice of ecological restoration. *Conservation Biology*, 25: 265–275.

Havens, K. and Vitt, P. (2016). The importance of phenological diversity in seed mixes for pollinator restoration. *Natural Areas Journal*, 36(4): 531–537.

Havens, K., Jolls, C. L, Marik, J. E., Vitt, P. and McEachern, A. K. (2012). Effects of a non-native biocontrol weevil, *Larinus planus*, and other emerging threats on populations of the federally threatened Pitcher's thistle (*Cirsium pitcheri*). *Biological Conservation*, 155: 202–211.

Havens, K., Vitt, P., Still, S., Kramer, A. T., Fant, J. B. and Schatz, K. (2015). Seed sourcing for restoration in an era of climate change. *Natural Areas Journal*, 35: 122–133.

Hereford, J. (2009). A quantitative survey of local adaptation and fitness trade-offs. *American Naturalist*, 173: 579–588.

Hipp, A. L., Larkin, D. J., Barak, R. S., Bowles, M. L., Cadotte, M. W., Jacobi, S. K., Lonsdorf, E., Scharenbroch, B. C., Williams, E. and Weiher, E. (2015). Phylogeny in the Service of Ecological Restoration. *American Journal of Botany*, 102: 647–648.

Hooper, D. U., Chapin, F. S., Ewel, J. J., Hector, A., Inchausti, P., Lavorel, S., Lawton, J. H., Lodge, D. M., Loreau, M., Naeem, S., Schmid, B., Setala, H., Symstad, A. J., Vandermeer, J. and Wardle, D. A. (2005). Effects of biodiversity on ecosystem functioning: A consensus of current knowledge. *Ecological Monographs*, 75: 3–35.

Hunter, M. (2007). Climate change and moving species: furthering the debate on assisted colonization. *Conservation Biology*, 21: 1356–1358.

Jefferson, L. V., Pennacchio, M. and Havens, K. (2014). *Plant-Derived Smoke and Seed Germination*. Oxford: Oxford University Press, 316 pp.

Kennedy, K., Albrecht, M. A., Guerrant, E. O., Dalrymple, S. A., Maschinski, J. and Haskins, K. E. (2012). Synthesis and Future Directions. In: Maschinski, J. and Haskins, K. E. (Eds). *Plant Reintroduction in a Changing Climate: Promises and Perils*. Washington, DC: Island Press, pp. 265–275.

Kildisheva, O., Erickson, T., Merritt, D. and Dixon, K. (2016). Setting the scene for dryland recovery: an overview and key findings from a workshop targeting seed-based restoration. *Restoration Ecology*, 24(52): 536–542.

Kramer, A. and Havens, K. (2009). Plant conservation genetics in a changing world. *Trends in Plant Science*, 14: 599–607.

Krauss, S. L. (2016). Seed sourcing for restoration of Banksia woodlands. In: Stevens, J. C., Newton, V. J., Barrett R. L., and Dixon K. W. et al. (Eds), *Restoring Perth's Banksia Woodlands*. Crawley, WA: The University of Western Australia Press.

Krauss, S. L. and Koch, K. E. (2004). Rapid genetic delineation of provenance for plant community restoration. *Journal of Applied Ecology*, 41: 1162–1173.

Krauss, S. L., Sinclair, E. A., Bussell, J. D. and Hobbs, R. J. (2013). An ecological genetic delineation of local seed-source provenance for ecological restoration. *Ecology and Evolution*, 3: 2138–2149.

Larkin, D. J., Steffen, J. F., Gentile, R. M. and Zirbel, C. R. (2013). Ecosystem changes following restoration of a buckthorn-invaded woodland. *Restoration Ecology*, 22: 89–97.

Leimu, R. and Fischer, M. (2008). A meta-analysis of local adaptation in plants. *PLoS ONE*, 3: e4010.

Madsen, M. D., Davies, K. W., Boyd, C. S., Kerby, J. D. and Svejcar, T. J. (2016). Emerging seed enhancement technologies for overcoming barriers to restoration. *Restoration Ecology*, 24(52): 577–584, doi: 10.1111/rec.12332.

Maschinski, J. and Haskins, K. E. (Eds) (2012). *Plant Reintroduction in a Changing Climate: Promises and Perils*. Washington, DC: Island Press.

Maschinski, J., Albrecht, M. A., Monks, L. and Haskins, K. E. (2012). Center for Plant Conservation Best Reintroduction Practice Guidelines. In: Maschinski, J. and Haskins, K. E. (Eds), *Plant Reintroduction in a Changing Climate: Promises and Perils*. Washington, DC: Island Press, pp. 277–306.

Maunder, M., Havens, K., Guerrant, E. O. and Falk, D. (2004a). *Ex situ* Methods: A Vital but Underused set of Conservation Resources. In: Guerrant, E. O., Havens, K. and Maunder, M. (Eds), Ex Situ *Plant Conservation: Supporting Species Survival in the Wild*. Washington, DC: Island Press, pp. 3–20.

Maunder, M., Havens, K., Guerrant, E. O. and Falk, D. (2004b). Realizing the full potential of *ex situ* contributions to global plant conservation. In: Guerrant, E. O., Havens, K. and Maunder, M. (Eds), *Ex Situ Plant Conservation: Supporting Species Survival in the Wild*, Washington, DC: Island Press, pp. 389–418.

McEachern, A. K., Bowles, M. L. and Pavlovic, N. B. (1994). A Metapopulation Approach to Pitcher's Thistle (*Cirsium pitcheri*) Recovery in Southern Lake Michigan Dunes. In: Bowles, M. L. and Whelan, C. J. (Eds), *Restoration of Endangered Species: Conceptual Issues, Planning and Implementation*. New York, NY: Cambridge University Press, pp. 194–218.

McHaffie, H. (2006). A reintroduction programme for *Woodsia ilvensis* (L.) R. Br. in Britain. *Botanical Journal of Scotland*, 58: 75–80.

McKay, J. K., Christian, C. E., Harrison, S. and Rice, K. J. (2005). 'How local is local?' – A review of practical and conceptual issues in the genetics of restoration. *Restoration Ecology*, 13: 432–440.

Merritt, D. J. and Dixon, K. W. (2011). Restoration seed banks: a matter of scale. *Science*, 332: 424–425.

Merritt, D., Martyn, A., Ainsley, P., Young, R., Seed, L., Thorpe, M., Hay, F., Commander, L., Shackelford, N., Offord, C., Dixon, K. and Probert, R. (2014). A continental-scale study of seed lifespan in experimental storage examining seed, plant, and environmental traits associated with longevity. *Biodiversity and Conservation*, 23: 1081–1104.

Miller, J. S., Lowry, P. P., Aronson, J., Blackmore, S., Havens, K. and Maschinski, J. (2016). Conserving biodiversity through ecological restoration: the potential contributions of botanical gardens and arboreta. *Candollea*, 71: 91–98.

Mounsey, C. (2014). Understanding Vegetation Patch-Gap Dynamics to Determine Restoration Success. PhD Thesis, University of Western Australia School of Plant Biology.

Naeem, S., Bunker, D. E., Hector, A., Loreau, M. and Perrings, C. (Eds) (2009). *Biodiversity, Ecosystem Functioning and Human Wellbeing*. Oxford: Oxford University Press.

Navarro-Cano, J. A., Ferrer-Gallego, P. P., Laguna, E., Ferrando, I., Goberna, M., Valiente-Banuet, A. and Verdu, M. (2016). Restoring phylogenetic diversity through facilitation. *Restoration Ecology*, 24(4): 449–455, doi: 10.1111/rec.12350.

Oldfield, S. and Newton, A. C. (2012). *Integrated Conservation of Tree Species by Botanic Gardens: A Reference Manual*. Richmond, UK: Botanic Gardens Conservation International, 55 pp.

PCA (2015). *National Seed Strategy for Rehabilitation and Restoration 2015–2020*. Washington, DC: Plant Conservation Alliance Federal Committee, Bureau of Land Management, 50 pp.

Ricciardi, A. and Simberloff, D. (2009). Assisted colonisation is not a viable conservation strategy. *Trends in Ecology and Evolution*, 24: 248–253.

Ritchie, A. L. and Krauss, S. L. (2012). A genetic assessment of ecological restoration success in *Banksia attenuata*. *Restoration Ecology*, 20: 441–449.

Rokich, D. P., Dixon, K. W., Sivasithamparam, K. and Meney, K. A. (2000). Topsoil handling and storage effects on woodland restoration in Western Australia. *Restoration Ecology*, 8: 196–208.

Rokich, D. P., Meney, K. A., Dixon, K. W. and Sivasithamparam, K. (2001). The impact of soil disturbance on root development in woodland communities in Western Australia. *Australian Journal of Botany*, 49: 169–183.

Rokich, D. P., Dixon, K. W., Sivasithamparam, K. and Meney, K. A. (2002). Smoke, mulch, and seed broadcasting effects on woodland restoration in Western Australia. *Restoration Ecology*, 10: 185–194.

SANBI (2010). *Monitoring Threatened Species in South Africa: A Review of the South African National Biodiversity Institutes' Threatened Species Programme: 2004–2009*. South African National Biodiversity Institute, 37 pp. Available online at: http://www.sanbi.org/sites/default/files/documents/documents/tspreview.pdf [accessed 25 April 2016].

Schwartz, M. W., Brigham, C. A., Hoeksema, J. D., Lyons, K. G., Mills, M. H. and van Mantgem, P. J. (2000). Linking biodiversity to ecosystem function: implications for conservation ecology. *Oecologia*, 122: 297–305.

SER (2004). *The SER International Primer on Ecological Restoration*. Society for Ecological Restoration International Science and Policy Working Group. Available online at: http://www.ser.org/resources/resources-detail-view/ser-international-primer-on-ecological-restoration [accessed 25 April 2016].

Sgro, C. M., Lowe, A. J. and Hoffmann, A. A. (2011). Building evolutionary resilience for conserving biodiversity under climate change. *Evolutionary Applications*, 4: 326–337.

Sharrock, S. L. (Ed.) (2012). *Global Strategy for Plant Conservation: a Guide to the GSPC. All the Targets, Objectives and Facts*. Richmond, UK: Botanic Gardens Conservation International.

Shaw, K. and Oldfield, S. (2013). *Enhancing Tree Conservation and Forest Restoration in Africa*. Richmond, UK: Botanic Gardens Conservation International, 59 pp.

Smith, A. B., Albrecht, M. A. and Hird, A. (2014). A plan for botanical gardens to facilitate movement of plants in response to climate change. *BG Journal* 11: 19–22.

Stevens, J. C., Rokich, D. P., Newton, V. J., Barrett, R. L. and Dixon, K. W. (Eds). (2016). *Banksia Woodlands: A Guide to their Restoration on the Swan Coastal Plain*. Crawley, WA: The University of Western Australia Press.

Tilman, D. (2001). Functional Diversity. In: Levin, S. A. (Ed.), *Encyclopedia of Biodiversity*, Vol. 3. New York: Academic Press, pp. 109–120.

Tilman, D., and Downing, J. A. (1994). Biodiversity and stability in grasslands. *Nature*, 367: 363–365.

Turner, S. R., Pearce, B., Rokich, D. P., Dunn, R. R., Merritt, D. J., Majer, J. D. and Dixon, K. W. (2006). Influence of polymer seed coatings, soil raking, and time of sowing on seedling performance in post-mining restoration. *Restoration Ecology*, 14: 267–277.

Turner, S. R., Steadman, K. J., Vlahos, S., Koch, J. M. and Dixon, K. W. (2013). Seed treatment optimizes benefits of seed bank storage for restoration-ready seeds: the feasibility of prestorage dormancy alleviation for mine-site revegetation. *Restoration Ecology*, 21: 186–192.

UN Climate Summit (2014). New York Declaration on Forests. Available online at: http://www.un.org/climatechange/summit/wp-content/uploads/sites/2/2014/07/New-York-Declaration-on-Forest-%E2%80%93-Action-Statement-and-Action-Plan.pdf [accessed 19 April 2016].

Vitt, P., Havens, K. and Hoegh-Guldberg, O. (2009). Assisted migration: part of an integrated conservation strategy. Letter in response to Ricciardi and Simberloff. *Trends in Ecology and Evolution*, 24: 473–474.

Wegener, M., Zedler, P., Herrick, B. and Zedler, J. (2008). Curtis Prairie: 75-year-old Restoration Research Site. Arboretum Leaflet 16. Available online at: http://www.botany.wisc.edu/zedler/images/Leaflet_16.pdf [accessed 19 April 2016].

Woodworth, P. (2013). *Our Once and Future Planet: Restoring the World in the Climate Change Century*. Chicago, IL: University of Chicago Press, 515 pp.

7 · *Botanic Gardens and Solutions to Global Challenges*

SAMUEL F. BROCKINGTON AND
BEVERLEY J. GLOVER

7.1 Introduction

The first two decades of the twenty-first century have been marked by a growing acceptance, both by politicians and by the public at large, of the need to focus on scientific research that can tackle the global problems facing humankind. In 2009 Professor John Beddington, then the UK government's chief scientific advisor, brought these issues to public attention by talking of a 'perfect storm' resulting from shortages of food, energy and water. Professor Beddington predicted that this storm would arise in 2030, based on data suggesting that global population would rise by 33 per cent, demand for water by 30 per cent and demand for food and energy by 50 per cent. It is both a challenge and a privilege for botanic gardens to be in a position to support scientific research that aims to mitigate some of these problems. Our aim in this chapter is to provide insight into how the maintenance, development and use of a living collection of diverse plant species can help in the development of solutions to some of the major problems facing the world in the twenty-first century. To tackle this question, we set out two key global challenges, those of global food security and global fuel security, and discuss the collection-driven research-oriented roles that botanic gardens can play in their solutions.

7.2 Global Food Security

For the last few years there has been a growing recognition that many of the solutions to the problems the world faces in feeding an ever-growing human population will lie in plant science. Global food security is thought to be at risk as a result of a combination of trends, including increases in population size and wealth, reductions in improvement in crop yields, the impact of climate change on land availability, and conflicts with demands to set land aside either to produce biofuels or to maintain biodiversity and store carbon.

The world's population continues to grow in size, demanding more food. The human population reached 6 billion in 2000, 7 billion in 2015, and is predicted to hit 8 billion by 2025 and 9–9.5 billion by 2050, after which population growth is expected to level off to some extent (UCSB, 2015). These extra people will clearly need extra food, but they may demand more extra food than simple numbers suggest. As populations grow in wealth, and particularly in flexibility of income, their use of meat as a primary foodstuff increases, and this creates an additional difficulty for global food supply. Since conversion of plant calories to animal calories has a maximum efficiency of 10 per cent, increased demand for meat creates at least a 10-fold increased demand on crop production (Godfray *et al.*, 2010). As of 2006, one-third of the world's total cereal production was used as animal feed, fuelling a 4.5-fold increase in the production of chickens and pigs over the previous 50 years (Godfray *et al.*, 2010). As these numbers increase, so demand for cereals will increase to keep pace. At present only four crops account for two-thirds of the calories consumed by the global population (Ray *et al.*, 2013). These four, maize, rice, wheat and soybean, have been the subject of extensive work to improve yield, and across the world the yield per unit area of land doubled for each of these four crops in the 50 years following 1960 (Ray *et al.*, 2013). These yield increases have been driven by the Green Revolution, and are the result of combined emphasis on agronomic improvements and technological improvements, coupled with intensive breeding to optimise yield within the genetic limits of each species. However, it is clear that yield increases are beginning to slow down, probably because the natural variation within key crop species has largely been exploited (Long *et al.*, 2015). For example, rice yield per hectare in China increased by 42 per cent in the 1980s, but only by 6 per cent between 2000 and 2010 (Long, 2014). This reduction in yield improvement is particularly problematic in light of the predicted increase in demand. Recent estimates converge on the idea that global demand for calories from crops will approximately double in the next 30–35 years (Tilman *et al.*, 2011; Ray *et al.*, 2012).

A further constraint on increasing crop yield comes from the availability of land on which to farm. Agricultural land availability has always been limited by soil type, elevation and climate, but other factors are beginning to have more prominence. There is increasing concern over the aridification and even desertification of land as a result of intense agriculture, and in some parts of the world, increased urbanisation also threatens arable land. The use of prime agricultural land for the production of biofuels, for example in the US, further limits the space available for food crops. As these difficulties increase, pressure mounts on areas of primary vegetation, which of course represent an important part of global diversity and an equally important carbon store to mitigate against climate change. Climate change itself threatens to further intensify all of these problems, potentially reducing land availability particularly through rising sea levels and increased desertification.

7.3 The Role of Botanic Gardens in Food Security

Botanic gardens have long played a role in maintaining collections for human health and nutrition, initially with roots in medicinal plants and physic gardens, and subsequently though the imperialism of economic botany (Brockway, 1979). Through the latter, networks of botanic gardens acted as a vital conduit and as agricultural research intermediaries to facilitate the establishment of plantation crops such as rubber (*Hevea brasiliensis*), quinine (*Cinchona calisaya*), tea (*Camellia sinensis*) and the breadfruit (*Artocarpus altilis*) (Brockway, 1979). But although the tradition of botanic gardens is rooted in the utilitarian, the contemporary role of botanic gardens has largely shifted in recent decades to one of horticultural display and biodiversity conservation. In turn, specialised agricultural research stations such as Rothamsted Research Station in the United Kingdom have long assumed the mantle of research into crop plants and plant breeding, now under the umbrella of food security (Russell, 1966). Botanic gardens, in turn, are perceived to contribute little to the goals of food security. However, recent surveys performed by BGCI reveal that botanic gardens can, and do, contribute in a number of ways to find solutions to the food security crisis (Sharrock, 2013).

In the BGCI survey, gardens were asked to rate the importance of various activities related to food security issues (Sharrock, 2013), of which working with local communities to enhance food production was rated an important part of their work by 80 per cent, conserving local crop varieties and crop wild relatives were rated important by 69 per cent, raising awareness of food security issues with the public was rated important by 65 per cent, providing seed and planting materials of food crops was rated important by 52 per cent, producing food crops in the garden for local use was rated important by 40 per cent, and, lastly, breeding new varieties of food crops was rated important by 35 per cent (Sharrock, 2013). Although only a fraction of botanic gardens were involved in breeding new varieties of crop plants, as we discuss, the phrasing of this question in the BGCI survey perhaps over-simplifies or narrowly defines the research contributions that botanic garden research can make to the area of food security. In the following sections we explore the role of botanic gardens in: (1) supporting collection-driven and trait-based approaches that underlie crop research; and (2) conserving and developing collections of landraces and crop wild relatives.

7.3.1 The Agronomic Demand for Systematic and Comparative Research

A major goal of research programmes operating within and around botanic gardens is to understand contemporary relationships among extant lineages of plants, and on the basis of these resolved relationships, to understand the patterns and processes of plant evolution. Although this research is rarely performed with the stated goal of crop breeding, fundamental systematic research can and does provoke evolutionarily informed research into crop improvement.

Biogeographic and phylogenetic analyses can identify the closest living relatives of crop plants, and identify centres of diversity for crop wild relatives, for subsequent use in breeding programmes. Analyses of natural diversity reveal that even complex agronomic traits have arisen multiple times across the flowering plants, and can indicate that the evolution of complex traits may be more simple than originally conceived, encouraging the possibility of genetic engineering in crop plants (Doebley, 1992). Separate origins of agronomic trait evolution can be examined for similarity in molecular convergence, potentially identifying primary targets for the engineering of these traits in crops, regardless of any constraints imposed by the distinct phylogenetic histories of different crop species. Finally, mapping the evolution of complex traits over a phylogeny can reveal step-wise transitions in the evolution of complex traits, providing simplified pathways for the accelerated artificial recapitulation of trait evolution by genetic engineering.

In the following sections, we explore how systematic research, of the kind driven by botanic gardens and their collections, is intellectually linked to food security. It is an important argument to make because the collection-based comparative biology that is spearheaded by botanic gardens is uniquely positioned to tackle many central questions in the arena of food security. A failure to recognise these opportunities limits the research scope of botanic gardens, and potentially limits progress in food security. In making this argument, we consider how fundamental systematic approaches are coupled to the genetic engineering of complex traits in food crops by looking at two case studies: the engineering of C_4 photosynthesis, and the goal of inducing root nodulation symbiosis in cereal plants. We then examine the importance of research into the origin and domestication of crop plants, and the role of plant systematics in the identification of the phylogenetic and biogeographic origin of wild crop relatives, using the example of the domestic apple.

7.3.2 Diversity-Driven Research into C_4 Photosynthesis

In photosynthetic plants, CO_2 is fixed by ribulose-1,5-bisphosphate carboxylase/oxygenase (RuBisCO) into C_3 carbon chains. However, RuBisCO also catalyses a competing oxygenation reaction, and the subsequent recovery of carbon through photorespiration is metabolically expensive. The term C_4 photosynthesis describes a suite of anatomical and biochemical adaptations that concentrate levels of CO_2 around the carboxylating enzyme RuBisCO, in order to improve photosynthetic efficiency in the context of high photorespiration rates (Sage, 2004). C_4 photosynthesis is far less common than the pervasive C_3 photosynthetic condition, but despite being present in only ~3 per cent of angiosperm species, it is estimated to be responsible for 20–30 per cent of terrestrial carbon-fixation (Sage et al., 2011). C_4 photosynthesis is a trait with a broad distribution across flowering plants, and is inferred to have originated in ~64 lineages distributed amongst 18 families of flowering plants (Sage, 2004). C_4 photosynthesis is also a remarkable example of convergent evolution, exhibiting considerable anatomical and biochemical variation among the different origins (Sage, 2004). Frequent convergence may suggest that C_4 evolution from a C_3 context is relatively easy, an observation that has encouraged efforts to engineer the C_4 pathway in C_3 cereal crops such as rice. A closer examination of the phylogenetic patterns of the C_4 pathway reveals highly localised clusters of C_4 photosynthesis. For example, most origins of C_4 photosynthesis are clustered within the monocot order Poales and the eudicot order Caryophyllales. Moreover, within these orders, multiple origins are further taxonomically restricted, with 18 origins in the Poaceae, and 15 origins within the Amaranthaceae in Caryophyllales (Sage, 2004). Clustering is most easily explained by variation in C_4 evolvability among taxonomic groups, and analyses of these clustered patterns with multiple origins are powerful tools with which to understand C_4 evolvability, and to inform efforts to artificially engineer the C_4 pathway in staple C_3 cereal crops.

All enzymes required for the C_4 pathway are present in C_3 plants, where they perform different, and often non-photosynthetic, functions. Similarly, the regulatory mechanisms necessary to target these genes to novel tissue locations in a C_4 context are also widely distributed across C_3 plants (Brown *et al*., 2011). From this background, C_4 genes have been iteratively recruited in the evolution of C_4 photosynthesis (Christin *et al*., 2013). Numerous analyses have sought to reconstruct the phylogeny of the C_4 pathway genes to understand the nature of C_4 gene recruitment. From these analyses, several findings emerge with significance to the *de novo* engineering of the C_4 pathway. Early reviews hypothesised a role for gene duplication in the evolution of the C_4 pathway, but subsequent phylogenetic analyses provide conflicting empirical evidence around this hypothesis (Hibberd *et al*., 2008). However, among C_4-related genes that belong to large multi-gene families, there is often asymmetry in recruitment. For example, across independent C_4 origins in Poaceae, the same PEPC lineage was repeatedly recruited out of the six possible subclades present in the grass genomes (Christin *et al*., 2007), and the same clade of NADP-malic enzymes was repeatedly recruited out of a possible four gene lineages present in grass genomes (Christin *et al*., 2015). Thus, certain gene lineages are more prone to recruitment, perhaps due to pre-adaptive catalytic properties, expression profiles or the evolvability of the gene in question. By extension, absence of these preferentially recruited gene lineages may therefore constrain the occurrence of C_4 photosynthesis, or constrain the type of C_4 biochemistry that evolves. However, in the subfamily of grasses Aristoideae, separate origins of C_4 photosynthesis are marked by the recruitment of distinct lineages of PEPC gene lineages. Phylogenetic analysis of selection patterns in C_4 genes commonly reveals that the same genes underwent parallel patterns of positive selection on key residues (Christin *et al*., 2007; Christin *et al*., 2009). For example, within the PEPC gene lineage in Poaceae, 21 codons are under positive selection in C_4 lineages, including that encoding the serine at site 780, which is essential for the catalytic activity of C_4-specific PEPC (Christin *et al*., 2007). Recurrent selection on these amino acids across polyphyletic origins of C_4 highlights the adaptive significance of these residues, and is relevant to the engineering of C_4 catalytic properties in proteins in a C_3 context.

The C_4 photosynthesis type is not a single discrete character, but a suite of features that function in concert. These features did not evolve simultaneously, but were acquired sequentially, as evidenced by several species of plants that are called C_3–C_4 intermediates (Christin et al., 2010). Taxonomic and systematic research has identified C_3–C_4 intermediates that possess only a subset of C_4, and in some instances are inferred to have subsequently given rise to full C_4 lineages, or in others have remained in varying degrees of C_3–C_4 intermediacy (Christin et al., 2010). Several groups have studied these intermediates, particularly in the daisy genus *Flaveria*, to identify important step-wise phases in C_4 evolution, as represented by the C_4 photosynthesis pyramid (Sage, 2004). Research on the natural pathways for C_4 evolution may provide important insights for overcoming the developmental barriers to C_4 photosynthesis (Sage, 2004).

7.3.3 Root Nodulation Symbiosis: From Evolution to Engineering

Fixed nitrogen is one of the limiting factors for plant growth, and a significant artificial input in modern agricultural systems (Charpentier and Oldroyd, 2010). A major biological source of fixed nitrogen (N_2) is derived from diazotrophic microorganisms living in the rhizosphere. One of the most efficient N_2-fixing systems for flowering plants is via nodulation symbiosis in which diazotrophic bacteria colonise the roots of host plants and provide fixed nitrogen in exchange for carbohydrates (Charpentier and Oldroyd, 2010). This form of endosymbiosis is restricted to four orders of flowering plants: Rosales, Fagales, Curcurbitales and Fabales. Traditionally these orders were thought to be phylogenetically disparate (Sprent, 2007), but molecular analyses identified them as closely related orders within a single group of flowering plants, suggesting the single origin of a predisposition to symbiotic nitrogen-fixation in angiosperms (Soltis et al., 1995). The restricted phylogenetic distribution of nodulation symbiosis, and their recent evolution (~105–70 Ma) (Doyle, 2011) long after the eudicot–monocot split, does not encourage the prospect of N_2-fixing symbiosis in monocot cereal crops. However, a closer look at phylogenetic patterns within the Eurosid 1 clade, and at the phylogenetic origins of genes underlying the evolution of N_2 symbiosis, is more encouraging with respect to engineering cereal crops. Although the predisposition to nodulation symbioses arose ~ 105 Ma, the origin of nodulation dates to ~70 Ma, thus the state of nodule predisposition must have been functional in non-nodulating lineages, maintained by purifying selection for over 30 Ma (Doyle, 2011). Identifying the molecular nature of this predisposition is an important research goal and will require a diversity informed and comparative sampling of both nodulating and non-nodulating taxa diverging prior to, and after, the hypothesised origin of this predisposition.

Following the evolution of this predisposition, nodulation has arisen as many as nine times separately within the Eurosid 1 clade (Doyle, 2011). These separate evolutionary origins have resulted in non-homologous nodules that can be determinate or indeterminate, be infected by root-hair dependent or root-hair independent infection processes, may or may not be associated with lateral roots, and can exhibit different biochemical pathways (Sprent, 2007). Trait variation among separate evolutionary origins of nodulation indicate that there are different evolutionary pathways to nodule symbiosis, which in turn may indicate a diversity of convergent mechanisms available for *de novo* genetic engineering (Charpentier and Oldroyd, 2010). Reconstruction of these nodule traits can in some instances reveal ancestral versus derived aspects of nodulation, and suggest potential first steps in the recapitulation of the evolutionary pathway by genetic engineering. For example, root-hair independent infection mechanisms appear to be the primitive state in legumes, do not require the same signalling mechanisms as root-hair infection, and may be a more realistic target for engineering rhizobial entry in non-legumes (Charpentier and Oldroyd, 2010; Doyle, 2011).

More importantly, across the many origins of nodulation, the nodulation processes have evolved from pre-existing plant developmental or symbiotic processes. For example, there are at least eight genes that function in both root nodule symbiosis, and in the more ancient arbuscular mycorrhizal (AM) symbiosis, suggesting that the evolution of root nodulation involved recruitment of the gene pathway from the more prevalent AM symbiosis (Charpentier and Oldroyd, 2010). Similarly, cytokinin and auxin signalling appear to act in concert to regulate the initiation of nodulation, but also regulate root developmental programmes such as lateral root formation and root meristem maintenance, so nodule organogenesis appears to have recruited pre-existing hormone signalling pathways (Charpentier and Oldroyd, 2010). These observations provide encouragement for engineering nitrogen-fixing monocots, as orthologous components of the machinery should already be present and functioning in cereal crops.

7.3.4 Determining the Relationships between Crops and Crop Wild Relatives

Crop wild relatives (CWR), the progenitors and close relatives of domesticated crop species, promise to be a potent tool for crop improvement. Domesticated species or crop plants typically include only a fraction of the genetic diversity found in their wild relatives, and genetic diversity within populations of CWR may harbour beneficial traits such as disease resistance or abiotic resistances such as salt tolerance. Effective breeding programmes involving domesticated crops and their wild relatives hinge to some extent on accurate identification of CWR, and on accurate analyses of their phylogenetic and biogeographic origins. The principal contribution of plant systematists and crop evolutionary biologists has therefore been in the identification of the wild progenitors of crop species. This is a challenging task, as the effect of strong artificial selection in the domestication process results in crops whose phenotypes are often considerably different morphologically from those of their relatives (Doebley, 1992). Key questions for plant systematists would then be: (1) What is the wild progenitor species? (2) When the progenitor is polymorphic and widespread, in what geographic region did domestication take place? (3) In the case of polyploid crops, what are the diploid ancestral species? (4) And, finally, has a particular crop species been domesticated once or many times? (Doebley, 1992). In answering these questions, plant systematics has the potential to address important questions for food security that underpin crop-breeding programmes. Here, we examine the role of these questions in the context of the domestic apple *Malus domestica*, a temperate fruit crop for which botanic gardens have played, and continue to play, a substantial role in understanding and conserving.

The apple is the most common and important fruit crop of northern temperate areas and its origin and domestication history is therefore of great interest. A suite of genetic studies has confirmed the wild Central Asian species *Malus sieversii* is the main contributor to the genome of the cultivated apple *M. domestica* (Cornille et al., 2014). More recently, it has been shown that a number of additional wild species have introgressed at some point in the domestication of *Malus*, including the wild European crab apple *M. sylvestris* (Cornille et al., 2012). Indeed, *M. sylvestris* is a major secondary contributor, with the result that *M. domestica* appears genetically more closely related to this species than to its Central Asian progenitor, *M. sieversii* (Cornille et al., 2012). However, a number of other species have contributed to a lesser degree to the *M. domestica* genome (Cornille *et al.*, 2014). Such inferences have important implications for breeding programmes as the contribution of the various wild species to the *M. domestica* gene pool highlights the need to invest efforts into the conservation of these donor species, which may contain unused genetic resources that could further improve the domesticated apple germplasm (Cornille *et al.*, 2014). Identifying the biogeographic areas in which these donor species exist is a critical component of any conservation initiative and breeding programme, and in this respect, morphological and molecular analyses have suggested that the wild origin of the apple was in Almaty, in Kazakhstan (Harris *et al.*, 2002).

In the vicinity of Almaty lies the Tien Shan mountain region, marked by elevational gradients, micro-climates and habitat niches, which make the region one of the world's most important centres of crop and plant genetic diversity, including for *Malus*. Environmental conditions in the region are so favourable that whole valleys are forested with apple trees, while varied ecological niches allow for evolutionary diversification of wild varieties. Yet the wild apple orchards of the Tien Shen mountains have been substantially destroyed through development and by hybridisation with domesticated apples (Cornille *et al.*, 2014). Here, the Institute of Botanics and Phytointroductions located in the Botanic Garden, Almaty, has played a vital role in the preservation and conservation of the wild apple, *Malus sieversii*. Indeed, the Institute and associated Botanic Garden have collected over 200 wild apple accessions, which are being preserved *ex situ* from forest locations, themselves also under *in situ* conservation practices. Over a period of about 40 years the apples have been gathered together to form an impressively diverse collection of materials, and are used to study the concentrations of biologically active substances present in the leaves and fruits, and to provide clonal materials for forest restoration. This case study emphasises how the combined analyses of crop evolutionary biologists together with local botanic garden-led research and conservation can drive the preservation of CWRs within the native area of origin, and promote their utilisation in plant breeding.

7.3.5 Conserving and Developing Crop Diversity, Landraces and Crop Wild Relatives

Undeniably, the most cost-effective method of providing plant genetic resources for long-term *ex situ* conservation is through the storage of seeds (Li and Pritchard, 2009). The main advantage of seed banking is that it allows large populations to be preserved (Schoen and Brown, 2001). Genetic erosion can be minimised by providing optimum conditions for storage, and furthermore, a considerable amount of diversity can be stored in a relatively small space (Li and Pritchard, 2009). In recognition of these advantages, the number of seed stores has dramatically increased over the last few decades, going from 54 in the 1970s to more than 1,300 in 1996 (FAO, 1996). Today, as a result of the continuous efforts of botanic gardens, and the focus on crop germplasm collections during the 1970s and the 1980s, approximately 6,000,000 accessions are contained in gene banks around the world. Indeed, seed storage accounts for about 90 per cent of the total accessions held *ex situ*, where an estimated 48 per cent of all accessions are advanced cultivars or breeders' lines, over 30 per cent are landraces or old cultivars, and about 15 per cent are wild or weedy plants or crop relatives (FAO, 1996). As of 1996, some 600,000 CWR accessions were maintained within a decentralised structure made up of 16 international agricultural research centres located in 12 developing and 3 developed countries. In this context, botanic gardens are unlikely to be majority holders of crop wild relatives, even though as many as 50 per cent of botanic gardens possess a seed bank, which contain many crop genera (Sharrock, 2013). However, botanic gardens can fill an important gap in *ex situ* conservation and have a comparative advantage in conserving vegetatively propagated plants, crops with recalcitrant seed, trees and other slow-growing species. Such species are frequently lacking in other *ex situ* germplasm collections.

A remarkable case study in the conservation and breeding of underutilised perennial crop species, led by a botanic garden, is the breadfruit (*Artocarpus altilis*). Breadfruit has long been an important staple crop and primary component of traditional agroforestry (Jones et al., 2011). The crop has many cultivars in production throughout the tropics, is rich in both carbohydrates and vitamins, and with all of the essential amino acids (Liu et al., 2015; Turi et al., 2015). The seeds germinate immediately and cannot be dried or stored, and are rarely used for propagation, thus breadfruit is usually propagated via root shoots or root cuttings. Consequently it is a crop that is ill-suited to standard agricultural seed banks, yet eminently suitable for maintenance in living collections in a botanic garden. The breadfruit collection located in the National Tropical Botanical Garden in Hawaii contains 226 accessions and approximately 120 varieties from 34 islands in the Pacific, with all three species of breadfruit and their natural hybrids (Zerega *et al.*, 2005). The National Tropical Botanical Garden not only houses the collection but also conducts research and agronomic improvement, including developing *in vitro* propagation methods, sensory evaluation of nutritional composition and fruit quality, documenting seasonal variations and productivity, and production of morphological descriptors to help identify and describe breadfruit varieties (Murch *et al.*, 2008; Jones *et al.*, 2010a,b). Fundamental research into agronomic improvement of breadfruit is coupled with an international programme that distributes cultivars of breadfruit to key tropical areas, such as in Zambia, Ghana and Nicaragua, to alleviate hunger, provide long-term security, and enhance the livelihoods of farmers.

7.4 Global Fuel Security

Global food security is a major part of Beddington's perfect storm, but is exacerbated by the accompanying issues of global fuel security. The rising demand from a growing world population for energy to fuel electricity production, transportation and industrial and technological infrastructure is a problem exacerbated by the finite supply of fossil fuels, the security of access to those fossil fuels and the environmental concerns arising from their use.

The world's growing population creates an inevitable demand for more fuel. However, as with global food supply, demands for fuel do not increase linearly with increasing population. Increased economic growth also drives increasing demand for fuel, as domestic electricity and the gadgets that consume it become more readily available, along with alternative forms of transportation. The predicted global population of 9 billion people in 2050 (UCSB, 2015) are likely to consume 50–56 per cent more energy than the 2010 global population, assuming 3–4 per cent global economic growth per year (Wagner *et al.*, 2016). Most of this increased demand for energy will be driven by developing countries, with about two-thirds coming from countries outside the Organisation for Economic Co-operation and Development (Leahy *et al.*, 2013).

This increased demand must be set against concerns about the amount of fossil fuel available and the longevity of the supply. In Europe, for example, over 75 per cent of energy used is derived from fossil fuels (Paschalidou *et al.*, 2016). While the debate about the supplies of fossil fuels is intense, and we are limited by lack of data on resources potentially available under deep-ocean, it is clear that fossil fuels will run out at some point. Coal is the most abundantly used fossil fuel, currently supporting around 50 per cent of US electricity production (IEA, 2006). At current rates of production coal supplies are predicted to last for another 150 years. However, if its use increases by 5 per cent per annum then supplies are predicted to dwindle by 2050 (IEA, 2006). Similarly, there is much debate about the likely timing of peak oil production – the maximum volume of oil extracted per day, after which rates of extraction and availability will decline. Oil currently accounts for one-third of global energy supply and 95 per cent of transport supply (Miller and Sorrell, 2014). A majority of estimates predict that peak oil extraction of over 90 million barrels per day could occur within the next decade, after which falling supply would soon fail to meet demand (Miller and Sorrell, 2014).

Economic and diplomatic issues arising from access to supplies of fossil fuels are likely to be of major global concern in the next few decades. In addition to these basic problems of supply and demand, the burning of fossil fuels also creates an environmental challenge, adding greenhouse gases to the atmosphere in an unsustainable way, and impacting on climate change. The UK, for instance, is committed to reducing its emissions of greenhouse gases by 60 per cent by 2050, a target that cannot be met with continued use of fossil fuel to the same extent (Pickett *et al.*, 2008). Solutions to this challenge, including the use of nuclear power, and a range of renewable energy resources, should also prove beneficial in the struggle to meet global energy demand.

One component of the renewable energy portfolio is plant-based biofuels, using photosynthesis to capture energy directly from sunlight and then converting it to useable hydrocarbon fuels. Biofuels are generally defined as liquid fuels derived from plant materials that can be used directly in engines, particularly to fuel transportation, in place of crude oil-derived products (Scott *et al.*, 2010). This ability to be used directly sets biofuels apart from other renewable sources of energy. They are also unique in that they are capable of being stored, something that is not easy with electricity-generating renewables such as solar or wind power. Biofuels may seem like a magic solution to global energy problems in an age of climate change, because the carbon they release when burned is only the same carbon that they fixed when growing – on first appearance they simply cycle carbon in an infinitely renewable way. However, all biofuels are beset by issues of input – if growth of the crop requires extensive agricultural input or processing of the biomass is not straightforward, then it is common for energy input into the crop to be at least 50 per cent of its energy output, reducing the efficiency and sustainability of biofuels considerably (Hill *et al.*, 2006). The growth of biofuels is of particular interest to plant scientists and botanic gardens, who must help to find a balance between demands to improve productivity and efficiency of biofuel crops with concerns over use of agricultural land or biodiversity-rich land for biofuel production. As with global food supply, climate change also impacts directly on the potential use of biofuels, changing habitats and opportunities to grow different crops in different areas. Central to understanding how to address this balance is an appreciation of the variety of biofuel options available. So before analysing how botanic garden-led research is driving innovation in fuel security, we will first outline the two main types of biofuels currently available – bioethanol and biodiesel.

7.4.1 Bioethanol

Bioethanol is the earliest form of biofuel (Sivakumar *et al.*, 2010). First generation bioethanol species were starch- or sucrose-based crop plants, already under agronomic production, namely *Zea mays* and *Saccharum officinarum*. However, the use of these food crops for fuel has a number of socio-economic consequences. In particular, the rising demand for bioethanol has competed with the traditional use of these crops as food and feedstock, and, exacerbated by government subsidies, has led to the displacement of food crop production to sensitive carbon-rich tropical forestlands. In turn, the loss of carbon-rich forests undermines one of the main values of plant biofuels, which is to offset CO_2 release. Studies suggest that extinction is occurring as a result of biofuel plantations in some regions, for example *Mitu mitu* (the Alagoas curassow) was native to the lowland forests on the Atlantic coast of Brazil, but clearing of forest land for biofuel sugarcane plantations has led to its extinction in the wild (Silveira *et al.*, 2004). Recognition of these issues has led to a focus on the use of native perennial crops that could be grown on marginal, abandoned and degraded land, and thus not compete with established food production systems (Heaton *et al.*, 2008). Furthermore, to avoid the use of valuable food crops and food substances, research and development has focused on developing second-generation bioethanol, in which the fuel is produced from lignin or cellulose, rather than from starch (Sivakumar *et al.*, 2010).

Lignocellulosic biomass represents a massive source of stored solar energy. Plants fix about 56 billion tons of CO_2 and produce over

170 billion tonnes of biomass annually, with cell walls representing about 70 per cent of that biomass, a substantial and hitherto untapped resource (Pauly and Keegstra, 2008). In this context two plant species, *Panicum virgatum* (switchgrass) and *Miscanthus giganteus* (giant miscanthus), have emerged as potential biofuel crops of the future. Both of these crops are perennial, and offer several advantages over annual crops because they require minimal inputs of pesticides and fertiliser, and are low maintenance (Sivakumar et al., 2010). Given our earlier discussion, it is interesting to note that both of these species exhibit C_4 photosynthesis, and are thus highly efficient at converting CO_2 into plant biomass for subsequent biofuels (Sivakumar et al., 2010). However, optimisation of plant-based biofuels from carbohydrate-rich lignocellulosic crops such as *Panicum* and *Miscanthus* requires the development of a number of novel processes. Breeding of selected ecotypes will be needed to maximise yields in particular climatic niches and biogeographic zones (Heaton et al., 2008). Here, *Miscanthus* already appears to be unusually cold-tolerant for a C_4 photosynthetic plant, and suitable for temperate conditions (Heaton et al., 2008). However, agronomic techniques will need to be developed to optimise the growth, culture and harvest of these crops. Furthermore, the breakdown of lignocellulosic biomass to ethanol requires the development and subsequent scaling up of biotechnologies capable of degrading these complex and variable polymer matrices (Sivakumar et al., 2010).

7.4.2 Biodiesel

An alternative to deriving biofuels from botanical biomass is to harness the oils naturally produced by plant species, which contain triacylglycerol fatty acid esters, each containing three fatty acid chains of C_{16}–C_{18} lengths (Sivakumar et al., 2010). These plant oils resemble conventional fossil fuels, and can be trans-esterified into biodiesel (Moser, 2009). Plant oils have more energy than ethanol-based biofuels, and only 10 per cent less energy than petroleum-derived diesel. In comparing biofuels and biodiesels, some estimates conclude that biodiesel from soybean has a 93 per cent net energy return, compared with only 25 per cent for maize derived bioethanol (Sivakumar et al., 2010). As with first generation bioethanol, the main crops adopted for biodiesel are the pre-existing food oil crops like rapeseed, sunflower and palm oil, and as with bioethanol, there are clear implications for land use. Studies have calculated that the UK transport industry used 47 billion litres of fuel in 2008,

53 per cent of it diesel. If that diesel were to be replaced with biodiesel, extracted from oilseed rape operating at its current optimum yield as a biofuel crop, 17.5 million hectares would be required to grow the plants (Scott et al., 2010). That represents approximately half of the entire land mass of the UK, a clearly unsustainable proposition. However, second-generation biodiesel crops offer a number of advantages in key areas. *Jatropha curcas* has received considerable attention recently for its potential as a biodiesel feedstock. As with perennial crops proposed for bioethanol, *J. curcas* appears to be suitable for marginal lands, requiring limited inputs (Sivakumar et al., 2010). The annual plant pennycress, *Thlaspi arvense*, is a further candidate for biodiesel production, with a seed-oil content of 20–38 per cent. The advantage of *T. arvense* is that it grows during late winter to early spring, thus providing a crop rotation option with another summer crop such as soybean (Johnson et al., 2007). Other non-food crops such as *Crambe cordifolia* and *Camelina sativa* are well adapted to growth in the mid-western and northern US, respectively. Other genera such as *Cuphea* produce shorter chain fatty acids of C_8–C_{14}, allowing oil to be used directly as biodiesel without trans-esterification, reducing post-harvest production costs (Johnson et al., 2007).

An alternative to terrestrial biofuels has arisen more recently, with the suggestion that algae might provide the same benefits as conventional crops but without the difficulties associated with land use. Algae are defined as aquatic photosynthetic organisms that are not embryophytes (land plants), and include a range of phylogenetically diverse unicellular and multicellular groups. Autotrophic and heterotrophic microalgae are an attractive option for the commercial production of biodiesel, as they often produce large quantities of triacylglycerides (TAGs) as lipid stores (Sivakumar et al., 2010). Some algae are able to accumulate very large amounts of TAGs. For example, *Ettlia oleoabundans* and *Nannochloropsis* sp. can increase their oil yield to about 50 per cent of dry mass when deprived of nitrogen (Gouveia and Oliveira, 2009). Such oleaginous algae are very attractive for use as a biofuel factory. An additional, and perhaps even more significant, benefit of using algae is that they can be grown in locations which are quite distinct from those used for food crop production. Since algae do not need soil, only light and water, it is possible to imagine a future where closed or open tanks of algae are situated on the roofs of buildings in urban environments or are grown on marginal or even extreme habitats where traditional crop cultivation would be impossible. However, algal biofuels present many of the same challenges in terms of optimisation and development that we see with terrestrial biofuels (Scott et al., 2010). Of the estimated 300,000 algal species, only a handful have been assessed as potential biofuels, and research is required to identify the best species for future breeding and development efforts. Input versus output concerns are an issue, with lipid deposition often occurring at the expense of growth (Scott et al., 2010). However, both are required for optimal yield, preferably with minimal nutrient input into the culture. Culture environments and processes are also the subject of much research, with closed algal bioreactors generating the best total yield but at energy costs than often outweigh the output. Open algal tanks potentially offer higher yield to input ratios but are heavily dependent on appropriate climate conditions, and can be disturbed by invasion of other microorganisms, which reduce yield (Borowitzka, 1999). Such problems of yield and cultivation are entirely analogous to the problems of developing any new crop plant, and will need to be addressed by the same combination of research, crop breeding and agronomic development.

7.4.3 Botanic Gardens and Biofuel Collections

The most obvious way in which botanic gardens can support the development of biofuels derived from terrestrial plants is by maintaining diverse living collections, and supporting collection-driven research. As with food crops, such collections provide opportunities for researchers to explore the potential of a range of species as novel crops. As is the case with food-based CWR, such collections also facilitate research on the wild relatives of biofuel crops with traits that might be of interest in expanding biofuel range, improving biofuel yield, or minimising biofuel crop input requirements. Access to potential biofuel species allows researchers to explore their traits, access to a range of varieties within a potential species allows comparison of key traits and analysis of the available diversity for use in breeding, and access to related species allows researchers to consider the opportunities for wide breeding or genetic manipulation, inspired by what traits are possible in a closely related genome. Alongside this key role in providing access to a diverse collection of plants, botanic gardens can also provide advice and insight into developing best horticultural and agricultural practice for novel and unusual crops in a range of environmental conditions. If the garden has sufficient experimental space, outside or under glass, to devote to small trials, then the horticultural staff can experiment with a range of inputs, growth conditions and time frames, and feed back into the biofuel development process.

7.4.4 *Jatropha curcus*: Developing a Novel Terrestrial Source of Biodiesel

A remarkable case study of a botanic garden that is driving collection-based research into biofuels, is Xishuangbanna Tropical Botanical Garden (XTBG) of the Chinese Academy of Sciences. Xishuangbanna Tropical Botanical Garden is the largest botanic garden in China, and its collections include over 12,000 species of tropical and subtropical plants cultivated in its 38 living collections. Of these, there are more than 5,400 energy plants of 350 species belonging to 50 families. Based on the compounds of biomass in the plants, the collection is divided into five functional sections, including oil, starch, hydrocarbon, fibre and firewood plants. These bioenergy plants are widely used or have a great potential for biofuel development in tropical and subtropical areas. The collection has fostered a broad range of applied and basic research into next generation biofuel plants, including work on the novel oil seed crop *Plukenetia volubilis*. But perhaps the most studied bioenergy plant in XTBG is the physic nut, *Jatropha curcus*. This species is a small tree or large shrub, up to 5–7 m tall, belonging to the Euphorbiaceae family (Kumar and Sharma, 2008). The plant has its native distributional range in Central and South America, but in cultivation has a pantropical distribution. Normally *J. curcus* flowers only once a year during the rainy season but in permanently humid regions or under irrigated conditions it flowers almost throughout the year (Kumar and Sharma, 2008). The plant is monoecious and the terminal inflorescences contain unisexual flowers. The ratio of male to female flowers ranges from 13:1 to 29:1 and decreases with the age of the plant (Kumar and Sharma, 2008). The female flowers produce seeds that contain viscous oil, and the species has emerged as a leading contender for biodiesel production due to the high level of oil in the seed, perennial life habit, and its suitability for cultivation on marginal agricultural lands (Sivakumar *et al.*, 2010). The XBTG is using its collections of *J. curcus* to sustain a significant research programme into the biology of *J. curcus*. Cumulatively, their approaches seek to establish basic biological experimental techniques in a poorly studied plant, and to manipulate resource allocation, in order to enhance female flowering and seed set. Significant research experiments into *J. curcus* have involved developing an *Agrobacterium*-mediated transformation system to enable genetic manipulation (Pan *et al.*, 2010), description of gene expression profiles at different developmental stages of growth of *J. curcus* (Chen *et al.*, 2011), analysing key proteins in the context of seed desiccation (Omar *et al.*, 2012), the use of exogenous giberellin to promote shoot branching (Ni *et al.*, 2015), and the isolation of key genes that specify reproductive identity (Li *et al.*, 2014). Such research will be essential for the rapid development of these crops in time to mitigate the challenges of climate change and fuel security.

7.4.5 Botanic Gardens and Algal Innovation

Although multicellular algae like *Chara* are reasonably common in botanic garden collections, it is also possible to grow unicellular species, providing the opportunity to develop a collection of microalgae for research purposes and to display that collection to the visiting public. Such a collection could provide the initial stocks for researchers exploring the lipidic content of different algae, their growth conditions, and the effect of different input regimes on their yield. However, it is not very likely that many botanic gardens will feel competent to develop a significant microalgae collection. Instead, their most useful role in the development of algal biofuels is likely to be in the provision of high quality growth space in which algal cultures can be grown and tested by researchers. Provision of space in polytunnels and glasshouses can allow researchers the opportunity to scale up algal growth from the millilitre scale to the scale of many litres, allowing more realistic analysis of biofuel potential. An example of this approach has been developed at the Cambridge University Botanic Garden, where a small-scale microalgae growth facility supported by a collaboration between the University of East Anglia and Cambridge University was opened in 2013. This algae facility was part of a network of pilot plants across NW Europe, where different algal species are being grown to establish what role algae can play in the development of a low carbon economy. The facility was based in a polytunnel, and contained growth space for multiple algae species in large polythene tubes through which oxygen was bubbled. The polytunnel itself ensured that the algae received sufficient light and heat. The facility also showcased a 6 m long photobioreactor with a capacity of 300 L. This photobioreactor was more energy-consuming, rotating, aerating, lighting and heating the culture automatically. It was designed on low energy principles to minimise input costs, and allowed analysis of input:yield ratios for different species. So successful was this small-scale facility that the Garden committed land and expertise in glasshouse design to the development of a much larger facility, the Cambridge Algal Innovation Centre, which opened in early 2016. The Centre is essentially a large glasshouse, with appropriate shading from some directions, that can be used to grow large-scale microalgal cultures at different temperatures and in different light conditions. It was built with a free-standing aluminium cladding system on a concrete foundation and has a gross floor area of 164 m^2. The aim of the project is that the Algal Innovation Centre will enable the scaling up of lab-based research and provide the pilot facilities that will allow researchers to test algal strains and growth conditions in a semi-industrial setting. The Centre provides the opportunity to connect the entire pipeline of algal research from strain selection and improvement, through harvesting and processing, to development of underpinning technology/engineering solutions. It is a prime example of lateral thinking in a botanic garden – although not a traditional use of space, in fact it provides an enormous boost to research aimed at solving major global problems using photosynthetic organisms.

7.5 Education and Public Awareness of Global Challenges

Botanic gardens, with their plant focus and predominant location in urban areas, must lie in the vanguard of efforts to educate the public on global challenges and their plant-oriented solutions. As briefly outlined here, there have been numerous notable and high profile education initiatives in this area, that range from education and public displays, to urban agriculture. Before exploring case studies, however, it is necessary to define some of the challenges inherent in botanical education and public outreach.

Botanists have long lamented the neglect of plants in all levels of the education hierarchy, from primary school textbooks to university lectures (Hoekstra, 2000). Students from primary school through to university can both better recall and name animals, and also prefer the study of animals to plants (Wandersee and Schussler, 1999). Such findings have led to accusations of zoo-chauvinism, which propagate the belief that studying animals is more important than studying plants. However, more recently, a number of authors have explored the concept of 'plant-blindness', which is defined as the inability to see or notice the plants in one's own environment (Wandersee and Schussler, 1999). This 'blindness' in turn leads to the inability to recognise the importance of plants in the biosphere and in human affairs; the failure to appreciate the biological features of plants; and the misguided, anthropocentric ranking of plants as inferior to animals, leading to the erroneous conclusion that they are unworthy of human consideration (Wandersee and Schussler, 1999). Two recent psychological experiments have found further support for both cognitive and visual perceptual components to plant-blindness, demonstrating that students have better recall of animal-based images (Schussler and Olzak, 2010), and are better able to detect animals than plants in rapid image sequences (Balas and Momsen, 2014).

Operating in tandem with visual-cognitive biases is the ignorance of the role that plants play in our food chain. Yet plant crops account for two-thirds of the calories consumed by the global population (Ray et al., 2013), and a third of the world's total cereal production is used as animal feed. Ignorance is not surprising given the processes of urbanisation, the divorce of rural and urban ways of living, and complex opaque food supply chains. Notable statistics in this regard include the observations that less than 2.5 per cent of the US population is directly involved in raising farm crops and that in the UK, 21 per cent of primary school children and 18 per cent of secondary school pupils have never visited a farm. It is perhaps not surprising, therefore, that children in particular seem ignorant about the role and origin of plant-based food products in our diet. For example, in a survey of 27,500 children, the British Nutrition Foundation (2013) found that 29 per cent of primary school children think cheese comes from plants, 10 per cent of secondary school children believe that tomatoes grow under the ground, and 34 per cent of 5–8 year olds and 17 per cent of 8–11 year olds believe that pasta comes from animals. These statistics highlight the significant challenges and opportunities for education programmes, emphasising the importance of public-facing plant-focused research institutes.

One of the most common ways in which botanic gardens are involved in helping with the global food security challenge is through education, with many institutions involved in growing food plants with local schools and community groups, providing horticultural advice for local food production, and providing a forum to debate food security issues. Often such events can be high profile, such as the IncrEDIBLES initiative at the Royal Botanic Gardens, Kew, which aimed to take visitors on a voyage through surprising edible plants. The garden-wide project ran for several months and involved a wide range of activities including talks and workshops, and online websites and blogs. Activities were accompanied by visual horticultural displays such as the '*Global Kitchen Garden*', showcasing over 90 edible plants from every corner of the globe, and '*The Tropical Larder*' which aimed to demonstrate how many daily foodstuffs have their origins in tropical rainforests. With reference to the inherent challenges of plant education, it is clearly the aim of these displays to highlight the global nature of the food supply chain, and the diversity of plants that make up our diet.

Education initiatives focused on the issues of global fuel security have been less common in the context of botanic gardens so far. However, an excellent example is the 'P2P' (Plants to Power) display in the Cambridge University Botanic Garden. The project is a demonstration of pioneering research into BioPhotoVoltaic and plant BioElectrochemical Systems (Bombelli et al., 2013). The display is essentially a power-generating hub in the form of a bus shelter. The sun shines on the P2P hub where solar panels use some of the light (mainly UV, blue and green) to generate negative and positive charges. These charges are then recombined through an external circuit to create electrical current. Living walls of plants on the side of the hub photosynthesise, grow and release organic compounds into the soil. These organic compounds feed bacteria living in the soil around the plant roots. The bacteria in the soil oxidise the organic compounds, producing electrons, which are collected by an anode, and protons, which are collected by a cathode. These electrons and protons are recombined and travel through an external circuit to produce further electrical current. The electrical current generated is monitored and displayed on a computer screen within the shelter. By integrating solar panels, living walls and plant BioElectrochemical Systems the P2P display demonstrates how solar energy can be used to generate electrical current by using and combining several different biological and non-biological technologies in a self-powering, sustainable system. The system has potential to supply energy in many environments where main supplies are unreliable. Well-lit bus shelters are an obvious use in urban environments, but the technology also has potential to provide essential electricity supplies in temporary housing situations, such as refugee camps. The P2P is a challenging display that excites the imagination, brings cutting edge research into the garden, and creates a clear synergy between technological ingenuity and horticultural innovation.

7.6 Conclusions

Food and fuel security are two global crises that will shape the world we live in over the twenty-first century. In this chapter we have explored how collection-based research and education programmes, driven by botanic gardens, can contribute to solutions to these societal challenges. In particular we have sought to demonstrate how collection-based comparative research has great potential to ameliorate the problems we face with respect to limited food and fuel supplies. We feel it is essential that gardens seek to direct their fundamental research base to tackle these challenges, and argue that comparative approaches are uniquely positioned to provide defined solutions in these arenas. Traditional fields of taxonomy and systematics, ever the strength of botanic gardens, have an explicit and foundational role to play in the development and breeding of novel crop traits and novel agronomic species for both food and fuel production. Furthermore, as public-facing research institutes, botanic gardens have an essential role in proselytising the central role of biodiversity in providing research-driven solutions. Here, our message must be that the goals of food and fuel security do not have to conflict with biodiversity conservation, but rather, that the solutions to these challenges, and the survival of global society, depend on the sustainable and ingenious use of biodiversity.

References

Balas, B. and Momsen, J. L. (2014). Attention 'blinks' differently for plants and animals. *CBE–Life Sciences Education*, 13: 437–443.

Bombelli, P., Iyer, D. M. R., Covshoff, S., McCormick, A. J., Yunus, K., Hibberd, J. M., Fisher, A. C. and Howe, C. J. (2013). Comparison of power output by rice (*Oryza sativa*) and an associated weed (*Echinochloa glabrescens*) in vascular plant bio-photovoltaic (VP-BPV) systems. *Applied Microbiology and Biotechnology*, 97: 429–438.

Borowitzka, M. (1999). Commercial production of microalgae: ponds, tanks, tubes and fermenters. *Journal of Biotechnology*, 70: 313–321.

British Nutrition Foundation (2013). National Pupil Survey 2013. UK Survey Results. www.nutrition.org.uk/attachments/608_uk%20Pupil%20Survey%20Results%202013.pdf

Brockway, L. H. (1979). Science and colonial expansion: the role of the British Royal Botanic Gardens. *American Ethnologist*, 6: 449–465.

Brown, N. J., Newell, C. A., Stanley, S., Chen, J. E., Perrin, A. J., Kajala, K. and Hibberd, J. M. (2011). Independent and parallel recruitment of preexisting mechanisms underlying C_4 photosynthesis. *Science*, 331: 1436–1439.

Charpentier, M. and Oldroyd, G. (2010). How close are we to nitrogen-fixing cereals? *Current Opinion in Plant Biology*, 13: 556–564.

Chen, M.-S., Wang, G.-J., Wang, R.-L., Wang, J., Song, S.-Q. and Xu, Z.-F. (2011). Analysis of expressed sequence tags from biodiesel plant *Jatropha curcas* embryos at different developmental stages. *Plant Science*, 181: 696–700.

Christin, P.-A., Salamin, N., Savolainen, V., Duvall, M. R. and Besnard, G. (2007). C_4 photosynthesis evolved in grasses via parallel adaptive genetic changes. *CURBIO*, 17: 1241–1247.

Christin, P.-A., Petitpierre, B., Salamin, N., Büchi, L. and Besnard, G. (2009). Evolution of C(4) phosphoenolpyruvate carboxykinase in grasses, from genotype to phenotype. *Molecular Biology and Evolution*, 26: 357–365.

Christin, P.-A., Freckleton, R. P. and Osborne, C. P. (2010). Can phylogenetics identify C(4) origins and reversals? *Trends in Ecology and Evolution*, 25: 403–409.

Christin, P.-A., Boxall, S. F., Gregory, R., Edwards, E. J., Hartwell, J. and Osborne, C. P. (2013). Parallel recruitment of multiple genes into C_4 photosynthesis. *Genome Biology and Evolution*, 5: 2174–2187.

Christin, P.-A., Arakaki, M., Osborne, C. P. and Edwards, E. J. (2015). Genetic enablers underlying the clustered evolutionary origins of C_4 photosynthesis in angiosperms. *Molecular Biology and Evolution*, 32: 410–858.

Cornille, A., Gladieux, P., Smulders, M. J. M. *et al.* (2012). New insight into the history of domesticated apple: secondary contribution of the European wild apple to the genome of cultivated varieties. *PLoS Genetics*, 8: e1002703.

Cornille, A., Giraud, T., Smulders, M. J. M., Roldán-Ruiz, I. and Gladieux, P. (2014). The domestication and evolutionary ecology of apples. *Trends in Genetics*, 30: 57–65.

Doebley, J. (1992). Molecular systematics and crop evolution. In: Soltis, P. S., Soltis, D. E. and Doyle, J. J. (Eds), *Molecular Systematics of Plants*. New York: Chapman and Hall, pp. 202–222.

Doyle, J. J. (2011). Phylogenetic perspectives on the origins of nodulation. *Molecular Plant–Microbe Interactions*, 24: 1289–1295.

FAO (1996). *The State of Food and Agriculture 1996*. FAO, Rome, Italy.

Godfray, H. C. J., Beddington, J. R., Crute, I. R. *et al.* (2010). Food security: the challenge of feeding 9 billion people. *Science*, 327: 812–818.

Gouveia, L. and Oliveira, A. C. (2009). Microalgae as a raw material for biofuels production. *Journal of Industrial Microbiology and Biotechnology*, 36: 269–274.

Harris, S. A., Robinson, J. P. and Juniper, B. E. (2002). Genetic clues to the origin of the apple. *Trends in Genetics*, 18: 426–430.

Heaton, E. A., Dohleman, F. G. and Long, S. P. (2008). Meeting US biofuel goals with less land: the potential of Miscanthus. *Global Change Biology*, 14: 2000–2014.

Hibberd, J. M., Sheehy, J. E. and Langdale, J. A. (2008). Using C_4 photosynthesis to increase the yield of rice–rationale and feasibility. *Current Opinion in Plant Biology*, 11: 228–231.

Hill, J., Nelson, E., Tilman, D., Polasky, S. and Tiffany, D. (2006). Environmental, economic, and energetic costs and benefits of biodiesel and ethanol biofuels. *Proceedings of the Natural Academy of Science, USA*, 103: 11206-11210.

Hoekstra, B. (2000). Plant blindness: the ultimate challenge to botanists. *The American Biology Teacher*, 62(2): 82–83.

International Energy Agency (2006). World Energy Outlook. Available online at: https://www.iea.org/publications/freepublications/publication/weo-2006.html [accessed March 2017].

Johnson, J., Coleman, M. D., Gesch, R. and Jaradat, A. (2007). Biomass-bioenergy crops in the United States: A changing paradigm. *The Americas Journal of Plant Science and Biotechnolog*, 1: 1–28.

Jones, A., Ragone, D. and Bernotas, D. W. (2011). Beyond the *Bounty*: breadfruit (*Artocarpus altilis*) for food security and novel foods in the 21st century. *Ethnobotany Research and Applications*, 9: 129–149.

Jones, A. M. P., Murch, S. J. and Ragone, D. (2010) Diversity of breadfruit (*Artocarpus altilis*, Moraceae) seasonality: a resource for year-round nutrition. *Economic Botany*, 64, 340–351.

Kumar, A. and Sharma, S. (2008). An evaluation of multipurpose oil seed crop for industrial uses (*Jatropha curcas* L.): a review. *Industrial Crops and Products*, 28: 1–10.

Leahy, M., Barden, J. L., Murphy, B. T. and Slater-Thompson, N. (2013). *International Energy Outlook 2013*. Washington, DC: US Department of Energy: The US Energy Information Administration.

Li, C., Luo, L., Fu, Q., Niu, L. and Xu, Z.-F. (2014). Isolation and functional characterization of JcFT, a FLOWERING LOCUS T (FT) homologous gene from the biofuel plant *Jatropha curcas*. *BMC Plant Biology*, 14: 125.

Li, D.-Z. and Pritchard, H. W. (2009). The science and economics of *ex situ* plant conservation. *Trends in Plant Science*, 14: 614–621.

Liu, Y., Ragone, D. and Murch, S. J. (2015). Breadfruit (*Artocarpus altilis*): a source of high-quality protein for food security and novel food products. *Amino Acids*, 47: 847–856.

Long, S. P. (2014). We need winners in the race to increase photosynthesis in rice, whether from conventional breeding, biotechnology or both. *Plant, Cell and Environment*, 37: 19–21.

Long, S. P., Marshall-Colon, A. and Zhu, X.-G. (2015). Meeting the global food demand of the future by engineering crop photosynthesis and yield potential. *Cell*, 161: 56–66.

Miller, R. G. and Sorrell, S. R. (2014). The future of oil supply. *Philosophical Transactions of the Royal Society A*, 372: 20130179.

Moser, B. R. (2009). Biodiesel production, properties, and feedstocks. *In Vitro Cellular and Developmental Biology – Plant*, 45: 229–266.

Murch, S. J., Ragone, D., Shi, W. L., Alan, A. R. and Saxena, P. K. (2008). In vitro conservation and sustained production of breadfruit (*Artocarpus altilis*, Moraceae): modern technologies for a traditional tropical crop. *Naturwissenschaften*, 95: 99–107.

Ni, J., Gao, C., Chen, M.-S., Pan, B.-Z., Ye, K. and Xu, Z.-F. (2015). Gibberellin promotes shoot branching in the perennial woody plant *Jatropha curcas*. *Plant and Cell Physiology*, 56: 1655–1666.

Omar, S. A., Elsheery, N. I., Kalaji, H. M., Xu, Z.-F., Song-Quan, S., Carpentier, R., Lee, C.-H. and Allakhverdiev, S. I. (2012). Dehydroascorbate reductase and glutathione reductase play an important role in scavenging hydrogen peroxide during natural and artificial dehydration of *Jatropha curcas* seeds. *Journal of Plant Biology*, 55: 469–480.

Pan, J., Fu, Q. and Xu, Z. F. (2010). Agrobacterium tumefaciens-mediated transformation of biofuel plant *Jatropha curcas* using kanamycin selection. *African Journal of Biotechnology*, 9: 6477–6481.

Pauly, M. and Keegstra, K. (2008). Cell wall carbohydrates and their modification as a resource for biofuels. *The Plant Journal*, 54: 559–568.

Paschalidou, A., Tsatiris, M. and Kitikidou, K. (2016). Energy crops for biofuel production or for food? Swot analysis (Case Study: Greece) *Renewable energy*, 93: 636-647.

Pickett, J., Anderson, D., Bowles, D., Bridgwater, T. and Jarvis, P. (2008). *Sustainable Biofuels: Prospects and Challenges*, London: The Royal Society.

Ray, D. K., Ramankutty, N., Mueller, N. D., West, P. C. and Foley, J. A. (2012). Recent patterns of crop yield growth and stagnation. *Nature Communications*, 3: 1293.

Ray, D. K., Mueller, N. D., West, P. C. and Foley, J. A. (2013). Yield trends are insufficient to double global crop production by 2050. *PLoS ONE*, 8, e66428.

Russell, Sir E. J. (1966). *A History of Agricultural Science in Great Britain, 1620–1954*. London: George Allen & Unwin Ltd.

Sage, R. F. (2004). The Evolution of C_4 Photosynthesis. *The New Phytologist*, 161: 341–370.

Sage, R. F., Christin, P. A. and Edwards, E. J. (2011). The C4 plant lineages of planet Earth. *Journal of Experimental Botany*, 62: 3155–3169.

Schoen, D. J. and Brown, A. (2001). The conservation of wild plant species in seed banks. *Bioscience*, 51: 960–966.

Schussler, E. E. and Olzak, L. A. (2010). It's not easy being green: student recall of plant and animal images. *Journal of Biological Education*, 42: 112–119.

Scott, S. A., Davey, M. P., Dennis, J. S., Horst, I., Howe, C. J., Lea-Smith, D. J. and Smith, A. G. (2010). Biodiesel from algae: challenges and prospects. *Current Opinion in Biotechnology*, 21: 277–286.

Sharrock, S. (2013). Botanic gardens and food security: the results of the BGCI's survey. *BGJournal*, 10, 2.

Silveira, L. F., Olmos, F. and Long, A. J. (2004). Taxonomy, history, and status of Alagoas curassow *Mitu mitu* (Linnaeus, 1766), the world's most threatened cracid. *Revista Brasileira de Ornitologia*, 12: 43–50.

Sivakumar, G., Vail, D. R., Xu, J., Burner, D. M., Lay, J. O., Ge, X. and Weathers, P. J. (2010). Bioethanol and biodiesel: alternative liquid fuels for future generations. *Engineering in Life Sciences*, 10: 8–18.

Soltis, D. E., Soltis, P. S., Morgan, D. R., Swensen, S. M., Mullin, B. C., Dowd, J. M. and Martin, P. G. (1995). Chloroplast gene sequence data suggest a single origin of the predisposition for symbiotic nitrogen fixation in angiosperms. *Proceedings of the National Academy of Sciences of the USA*, 92: 2647–2651.

Sprent, J. I. (2007). Evolving ideas of legume evolution and diversity: a taxonomic perspective on the occurrence of nodulation. *New Phytologist*, 174: 11–25.

Tilman, D., Balzer, C., Hill, J. and Befort, B. L. (2011). Global food demand and the sustainable intensification of agriculture. *Proceedings of the National Academy of Sciences*, 108: 20260–20264.

Turi, C. E., Liu, Y., Ragone, D. and Murch, S. J. (2015). Breadfruit (*Artocarpus altilis* and hybrids): A traditional crop with the potential to prevent hunger and mitigate diabetes in Oceania. *Trends in Food Science and Technology*, 45: 264–272.

USBC (2015). World population 1950–2050. International Database (Washington, DC: U.S. Department of Commerce, Census Bureau).

Wagner, L., Ross, I., Foster, J. and Hankamer, B. (2016). Trading Off Global Fuel Supply, CO_2 Emissions and Sustainable Development. *PLoS ONE*, 11(3), e0149406.

Wandersee, J. H. and Schussler, E. E. (1999). Preventing plant blindness. *The American Biology Teacher*, 61: 82–86.

Zerega, N. J. C., Ragone, D. and Motley, T. J. (2005). Systematics and species limits of breadfruit (*Artocarpus*, Moraceae). *Systematic Botany*, 30, 603–615.

8 · Cultivating the Power of Plants to Sustain and Enrich Life
How Public Gardens Can Realise our Purpose by Focusing on the Basic Human Needs Universal to Diverse Audiences

SOPHIA SHAW AND JENNIFER SCHWARZ BALLARD

This chapter is dedicated to explaining how botanic gardens can increase the diversity of their visitors and programme participants and intensify visitors' level of involvement. There are many successful case studies from gardens around the world that demonstrate how gardens can engage people from all ages, backgrounds and abilities in all we do; we will highlight some of these.

The need to engage the broadest and most diverse audience in garden activities has never been more urgent. Plants, as a result of climate change, habitat fragmentation, invasive species, and other factors, are in peril. All life depends on plants – we rely on them for our food, clean air and water, medicine, shelter and clothing – and healthy ecosystems, all of which have plants at their core. By including everyone in our passion for plants and inspiring in all people an appreciation of nature, we can best fulfil our mission as botanic gardens. When we increase people's understanding of the importance of plants and plant diversity, we help build a citizenry of environmentally conscious civic leaders, scientists and stewards of our planet.

However, engaging a broad audience and inspiring all people to be environmentally conscious is an ambitious goal. We still have far to go to reach a day when botanic garden audiences reflect the diverse demographics of our regions. We have history to overcome: most American and many European gardens were created from the private estates of white, upper-class individuals or through their personal philanthropy, and organised in ways that served the pleasures, social desires and intellectual curiosities of a white educated upper class. Historically, public gardens have been located in affluent neighbourhoods, and even if those neighbourhoods are diverse, visitorship has tended to consist of mostly white, generally affluent people. Often unintentionally, the programming, marketing, or cost of entry kept people whose backgrounds were different from the gardens' founders outside the garden gates. Changing this is both a matter of becoming more welcoming and intentionally altering traditional garden culture.

How we overcome this history and successfully achieve a more diverse visitorship is a question garden and museum leaders have debated and advanced passionately since at least the 1970s. Efforts to date have been guided primarily by the term 'community outreach' – reaching out from the institution's main building or garden site to people, most typically those who are poor and not white. The efforts have resulted in off-campus programmes in schools, parks and community centres; displays and exhibitions intending to appeal to varied interests; materials and tours translated into different languages; education and training programmes; and new on-site and online marketing tools that make an effort to anticipate the needs of diverse audiences. The invention of 'community outreach' was transformational in this effort of gardens and museums – allowing for important progress in changing the population of people served. However, despite garden and museum leaders' sincere commitment to these efforts, they were misguided in that they often stereotyped the needs and interests of new audiences, making assumptions about what would appeal to a specific demographic; for example, programming African drummers to attract black Chicagoans or offering Mexican food to attract local Latinos – neither of which results in a significant visitor demographic shift. Despite shortcomings, however, these efforts certainly helped by creating familiarity with a previously unknown institution.

However, this approach is insufficient to fulfil botanic gardens' mission today. Both the semantics and concept of 'community outreach' have served out their useful life. We should now do away with this inside-out framework and the words that describe it if we are to achieve our goal of truly serving and engaging diverse audiences.

We must do two things. First, we must enhance garden programmes that serve people's basic contemporary human needs, recognising we all share dreams and challenges, regardless of where we live, how much money we make, our age, skin colour, gender or other superficial differences. We all need access to food, education and exercise. We all need support when we fall sick or grieve, and as we age. We all need safe places where we can relax, celebrate and spend our leisure time. The quality of all our lives depends upon a healthy future for our planet. Second, we must *authentically* engage the audiences we seek to reach, whether on site or in a neighbourhood. This requires that we interact on equal footing. We must be willing not only to talk but also to listen, and respond to diverse interests, desires and concerns through our actions and programmes. Botanic gardens, because of the deep human connection with nature, have a unique opportunity to do this.

8.1 Reframing 'Community Outreach' as Authentic Engagement

In order to identify a framework for engaging diverse audiences and specific tools to guide that approach, we will first deconstruct the concept of 'community outreach' and then suggest an alternative way of thinking about the services gardens provide.

According to the (online) *Oxford English Dictionary*, 'outreach' is defined as, 'An organisation's involvement with or activity in the community, especially in the context of social welfare.' In the garden or museum context this reaching out has generally come in the form of a monologue rather than a dialogue, where an institution makes a priori judgements about its audience. Perhaps this explains why many museum and garden outreach efforts have not worked; they have been inadvertently grounded in unequal power dynamics and untested assumptions about the interests and needs of our target audiences.

And then there is 'community' – also a word that is problematic in the context of audience engagement. We are once again forced to reflect upon the ways language shapes both action and reaction. Community suggests more than commonality, it implies a shared interest, culture or context that defines individuals as a group. This can take place internally, by the members themselves, or more problematically, externally, in which case it can become little more than a stereotype. For example, journalists reporting on how different groups of individuals will vote in an election may characterise the 'Latino community' as supporting immigration reform or 'women voters' supporting Hillary Clinton for US president. However, we know that people do not act or vote in homogeneous blocks and that we each can identify with multiple 'communities' at the same time. As cultural institution administrators, thinking along the lines of 'appealing to a community' can often lead to programme decisions (certain cultural exhibitions or festivals for example) that can inadvertently be understood as biased and contributing to stereotypes, thus distancing the very people whom we hoped to engage. At the same time, we cannot let the diversity of our audience overwhelm us – it is better to try, and to recognise publically and explicitly the multifaceted nature of any group, than not to engage with new audiences at all.

The words 'community' and 'outreach' were chosen with good intention, but the terminology has hindered deeper, reciprocal conversations with potential audiences. So, let us abandon the language of 'reaching out' to a 'community'. Rather, let us use language that recognises the commonalities among all people and celebrates the differences: one that starts with openness and a willingness to listen and a desire to understand, one that draws in people from all ages, backgrounds and abilities to botanic gardens and museums. Let us reframe this process as authentic engagement, as conversation among equals. We will not broaden or diversify our audience by thinking that people who speak a language other than English, or who have less money, education or mobility cannot, metaphorically, step over the fallen tree or climb the hill on their own.

Garden leaders must aim to serve basic human needs in order to expand audiences and increase impact. Replacing the out-dated concept of the erudite institution attempting to reach out is a way of thinking that encourages us to ask, 'how do we serve the individual, while meeting the universal needs of all people?' In other words, how can we both ensure our garden is relevant to our users, visitors or customers, while at the same time be conveners of people of different backgrounds coming together at an exciting moment of opportunity to bring about positive social impacts against a backdrop of news that often seeks to divide us. Let us celebrate the fact that our living museums have the capacity, because of our commitment to nature, to address a diverse suite of needs; both those of our planet and of its diverse and complex human population. Garden grounds and programmes build people's social, physical, emotional and intellectual strength in significant ways every day. What an exciting and important time, full of opportunity, it is to be working in places with so much potential to do good.

But it is not just about doing good.

Institutions that continue in the old model of community outreach may not thrive to their potential. Instead, we must with profound commitment, invite people in. We must put human beings, not the institution, in the centre. We must think about the 'customer' (and we use this term consciously, because in traditional business vernacular the customer is 'always right') and identify how our garden can serve them, not by making assumptions, but through dialogue and conversation. We must put the needs of our stakeholders, whether they are visitors, students, donors, volunteers, employees, vendors or programme partners, at the centre of our plans. If we focus on truly serving people of all backgrounds, abilities and ages – from birth through to death – we will thrive, and so will our message about plants.

8.2 How Can Botanic Gardens Broaden their Reach and Make a Difference?

Now that celebrating diversity and adopting authentic engagement are our framework, how then do we apply these in operations and programming to reach our current and desired audiences? Specifically, what do we do?

At the most basic level, garden presidents, together with their boards of directors and executive staff, must decide to make diversity a strategic priority and make decisions at every level of the organisation that support it. They must commit, not only verbally, but to the core of their belief system, that attracting people of all backgrounds to their garden is beneficial and the right thing to do. They must create a culture of acceptance within those groups who feel their needs are not being met are comfortable voicing their concerns with the confidence they will be taken seriously. Even today, there are those who would prefer not to hear other languages, or sit at a café table with people with whose skin pigments or clothes do not match their own. If these attitudes exist at the leadership level of an institution (board and staff), they must be addressed directly and openly, if necessary with the help of an outside mediator. The growth and positive change achieved at the Chicago Botanic Garden are in large part a direct result of a leadership-level staff and board commitment and partnership, one also embraced and forwarded by our government partners, the Forest Preserves of Cook County.

Then, we realise that we must review and assess every aspect of our institutions' policies, programmes and communication vehicles, looking critically for inherent bias or exclusionary language or symbols. We must ask ourselves honestly, 'why are we not attracting a more diverse audience, workforce, board, etc.' and speak openly about what changes we can make both for the short and long term. After a decade of effort at the Chicago Botanic Garden, we made many small changes to policies and programmes that might seem individually insignificant, but which together create a rich, consistent, reliable structure that allows for operational, programmatic and individual growth. From mission statement and personnel policies, to purchasing, programming and marketing, equality and openness must guide action.

At a basic practical level, it means considering how organisational structures affect staff. Of its nearly 300 permanent full-time employees, the Chicago Botanic Garden employs approximately 60 staff members for whom Spanish is their only or preferred language. In peak season, we add 50 Spanish-language-preferred temporary workers to this number. There are also employees who are unable to write or read in either English or Spanish. Yet until 2011, there was no one on the management team who could communicate with this essential workforce. In response, the Garden included fluency in Spanish in the hiring criteria for the new vice president of human resources. The culture of the institution started to shift immediately. Spanish-speaking employees were finally offered the right health-care benefit plans for their families; individuals struggling at work or with a personal crisis could find the help they needed. Lesson learned: you cannot build a positive institutional culture, help people to excel or, much less, diversify your team, if you cannot communicate with everyone.

Likewise, we reviewed our policies regarding same-sex couples and were surprised to find same-sex partner benefits did not mirror those offered to heterosexual staff members. Upon further research, we learned that in our successful event business, staff had been instructed not to book events celebrating the union of same-sex partners. Both of these policies were changed immediately. But it was against this background that our senior staff, with support of the board's executive committee, decided to participate in the Pride Parade in Chicago – attended by almost a million people. Our horticultural staff created a colourful float from our trolley. Staff and volunteers danced and walked for hours along the parade route. After the first year of our participation, four Garden employees made a point of saying how much our participation meant to them. One noted that she finally 'felt like she belonged' at the Garden. The Garden's presence at the parade made a public statement of acceptance and respect to the diverse crowd of parade-goers and those who watched on television.

Operationally, inclusion extends beyond staff to those with whom we do business: our contractors and suppliers. The Supplier Diversity Programme, developed with our partners the Forest Preserves of Cook County and the Chicago Zoological Society's Brookfield Zoo, aims to ensure that the Garden actively seeks out women- and minority-owned businesses to provide services and products. The bidding process requires that staff publicise opportunities to diverse audiences, and consider diversity when awarding contracts. Contractors, like staff, become part of the Garden 'family' for a time, to learn about the Garden's mission and purpose. They also, upon completion of a project, feel invested in the campus and perhaps become more likely to visit or share the experience and purpose of the Garden with family and friends. If we expand our list of experienced suppliers, we naturally expand the understanding of our mission and reach of our institution. Botanic gardens can achieve operational sustainability and build understanding of plant conservation only if everyone – from diverse neighbourhoods and backgrounds – has the opportunity to become involved in our institutions. In Chicago, we have seen, and felt, results.

Finally, this means creating a culture in which all stakeholders are comfortable in voicing their unmet needs. For example, the neighbourhoods surrounding the Chicago Botanic Garden have a high Jewish population. In the Jewish tradition, the late-winter holiday Tu B'Shevat is a time for ushering in the 'new year for trees'. The day is currently celebrated as an ecological awareness day and a welcoming of spring. In response to this specific, culturally relevant, demand, we developed programming for field trips and families that aligned with Tu B'Shevat that are now popular with all our audiences.

To recognise and reinforce our commitment to diversity, in late 2014, the staff, board and volunteers of the Chicago Botanic Garden adopted a new mission statement: 'We cultivate the power of plants to sustain and enrich life.' We wrote this not only to underscore the conservation message of the importance of plants, but also to include all of the people who help the Garden to fulfil its goals. The previous mission statement was descriptive, yet oblique: 'to promote the enjoyment, understanding, and conservation of plants and the natural world'. Who was doing the promoting? Why? For whose benefit? Yet, by emphasising three words: 'we', 'cultivate' and 'enrich', we provided clarity. The words 'We cultivate' include everyone – from garden horticulturists to course instructors (who cultivate the minds of students) to ex-offenders who may be enrolled in our Windy City Harvest urban agriculture programmes. 'Sustain' refers to the fact that all life depends on plants. And 'enrich' gives merit to the mental benefits of our work, acknowledging the commitment of our horticultural therapists and yoga instructors, as well as emphasising the importance of the beauty, community, inspiration, joy and comfort that botanic gardens provide.

Only with everyone in the organisation working together, using the best of our diverse perspectives and backgrounds, can we make progress in achieving our bigger ambitions to save the planet by saving plants; identifying methods to mitigate the impact of climate change; and using the power of nature to beautify and restore neighbourhoods, and achieve social justice.

8.3 Conclusion: Change Won't Come Easily, but the Stakes Are High and the Benefits Are Well Worth the Effort

Botanic garden leaders must be agents of change if we are going to fulfil our mission. This is not always easy, and is occasionally risky, but people and plants depend on us. In Chicago, when we're not sure what to do, we review the issue at hand through the lens of what is best for our stakeholder – 'what would best meet his or her needs?' We also remember the importance of our purpose and review the following value statements, on which our strategic plan is built:

- We believe: beautiful gardens and natural environments are fundamentally important to the mental and physical well-being of all people.
- We believe: people live better, healthier lives when they can create, care for, and enjoy gardens.
- We believe: the future of life on Earth depends on how well we understand, value and protect plants, other wildlife, and the natural habitats that sustain our world.

In conclusion, gardens will become more operationally sustainable and financially viable if we lead as agents of social justice. As garden leaders and enthusiasts, we must use the unique power of plants to sustain and enrich the lives of our stakeholders – all people of every age, background and ability – in the ways they choose, grounded in satisfying the basic human needs we all share. When those we engage thrive, we thrive. The more diverse our audiences become, the more broadly our education and conservation missions will be shared around the globe – allowing more people to become engaged with our programmes and to live better, happier, healthier, more satisfying lives. Gardens can support people facing many challenges. Gardens are a refuge. Our campuses – whether big or small, urban or suburban – provide comfort and present solutions in science education, food security, workforce training, wellness and climate change. We offer solace to people facing crisis, and a place to celebrate or find joy. Gardens will remain sustainable – and we firmly believe they will – if we see the world's greatest problems as the world's greatest opportunities to serve. The more relevant our connections to our audiences, the more successful our institutions will be. Presented here are four case studies illustrating these concepts and principles.

Figure 8.1 Windy City Harvest Youth Farm students selling at a local farmers' market. For the colour version, please refer to the plate section. In some formats this figure will only appear in black and white.

Case Study 8.1 *We Use Nature-Based Solutions for Social Justice: Windy City Harvest*

By Angela Mason, Chicago Botanic Garden

Through youth development, workforce training and job placement, entrepreneurship and farm business development, the Chicago Botanic Garden's Windy City Harvest's urban agriculture programmes empower at-risk youth and adults, including those involved with the justice system, to find a pathway to a successful future.

A four-tier training programme comprising Youth Farm, Apprenticeship, Harvest Corps and Entrepreneurship and Careers supports workforce development while building a local food system, healthier neighbourhoods and a greener economy. The produce grown and harvested at the 13 community-based sites reaches populations the traditional US food system has bypassed, especially low-income communities where grocery stores are scarce and rates of diet-related diseases are high. The work-related training increases the economic security of participants and communities.

Youth Farm

Each year, the Windy City Harvest Youth Farm educates and employs 80 to 90 teens from low-income communities at three farm sites in Chicago and one in Lake County, IL. As they advance through a programme grounded in sustainable urban agriculture and social emotional learning principles, Youth Farm students learn to grow food responsibly, work as a team, advocate for food justice, eat in a healthy way and become accountable – to themselves, their fellow farmers and to their employers. Through nutrition demonstrations and redemption of federal nutrition assistance coupons at markets and Women, Infants and Children (WIC) offices, Youth Farm students engage their communities and gain a better understanding of pressing social and economic issues.

Youth Farm teens work in all aspects of sustainable farming and food systems – from planting a farm to managing a beehive, from cooking with the food they grow to selling it at local farm stands and markets, and through sales to the Garden View Café, where the chef incorporates fresh organic produce into menu items available to Chicago Botanic Garden visitors. Teens are paid a stipend for four hours per week in the spring and fall, and 20 hours per week in the summer, but the benefits far outweigh the wages they earn. By the end of the season, they have gained valuable job and teamwork skills, discovered a whole new way to look at the food they eat, and grown their support system to include supervisors, programme coordinators, legislators, and their fellow participants. The personal and professional impacts carry far beyond the programme end. On average, over 80 per cent of Youth Farm participants enrol in post-secondary education (compared to 29 per cent of all students entering Chicago Public High Schools). As North Lawndale student Anton Willis remarked, 'I know this job is going to help me to get another job.'

Apprenticeship

Through the nine-month apprenticeship programme adults earn a sustainable urban agriculture certificate accredited by the Illinois Community College Board. Students begin with coursework. Then, from June through September, they complete internships at Windy City Harvest and partner farm sites. In September, students return to the classroom for sessions devoted to job searching, résumé building and mock interviews.

The experiences of recent graduates are evidence of the programme's success; as of 2015, 91 per cent of graduates have started their own businesses, are employed as beginning farmers or within the local food sector. Their positions include the rooftop gardening coordinator of Revolution Brewing, manager of the edible landscaping division at Christy Webber, GreenCorps mentor at Tilden High School, Windy City Harvest sales coordinator, and the manager of Mariposa Gardening and Design – who is now enrolled in a sustainable landscape architecture programme at the University of California, Berkeley, and interned at several French chateau gardens, including the Palace of Versailles, as part of the Garden's exchange programme with the French Heritage Society.

Harvest Corps

Working with the Illinois Department of Juvenile Justice (IDJJ) Youth Centers as well as other organisations supporting at-risk young people, Windy City Harvest Corps offers educational and transitional jobs programmes for justice-involved youth aged 17–21, and others who have significant barriers to employment. The Corps is a modified 13-week apprenticeship programme, with training in environmental literacy and work skills through the Roots of Success curriculum (which engages students by making learning relevant, building on prior knowledge and experience, and connecting education to employment and further learning in green career fields. This is supplemented with additional support service coordination (e.g. for housing or food stamps) and closely supervised work assignments to ensure that as these individuals transition back into their communities, they have the support systems necessary to be successful.

Figure 8.2 Corps participant harvests lettuce on the McCormick Place rooftop in Chicago, IL. For the colour version, please refer to the plate section. In some formats this figure will only appear in black and white.

Entrepreneurship and Careers

Launched in 2013, the incubator farms of Windy City Harvest support entrepreneurs as they start or manage farms and farm-related businesses. Four industry-specific certificates are also offered: Business and Entrepreneurship for Local Foods, Season Extension, Aquaponics and Vertical Farming Systems, and Rooftop Farming and Edible Landscaping.

With support and mentoring from Windy City Harvest staff, students develop business plans and launch successful urban farm businesses and ventures. One example is Your Bountiful Harvest Family Farm, developed by Safia Rashid, who grows vegetables for WIC boxes, the Legends South Community and wholesale distributors, including Midwest Foods. Rashid worked her incubator farm plot with the help of her husband, volunteers and a team of interns. The 39-year-old mother of three grew vegetables to fill boxes for the WIC programme. 'It makes me very happy,' said Rashid, a onetime WIC recipient, 'I feel like I'm coming full circle to give something back to a programme that gave to me.'

Impacts

Since its inception in 2003, the Windy City Harvest has impacted individuals and communities:

- 900 high school students from Chicago's low-income communities participated in the Youth Farm;
- 180 formerly incarcerated individuals participated in the Harvest Corps transitional jobs programme;
- 109 adults completed the apprenticeship programme;
- 22 adults completed the Business and Entrepreneurship for Local Foods and Season Extension certificate courses;
- 12 beginning farmers were mentored and supported in the creation of six urban farm businesses;
- 525,795 pounds of fresh fruits and vegetables were sustainably grown and harvested, generating over $1 million in revenue and serving more than 875,000 low-income community members;
- 10,653 boxes of produce were distributed to Community Economic Development Association of Cook County (CEDA) Women, Infants, and Children (WIC) clients in exchange for food vouchers;
- Over 80,564 pounds of produce were donated to food pantries and community health centres, such as the Pacific Garden Mission and Pilsen Wellness Centre, or taken home by students and staff.

Stacy Kimmons, a 2014 Windy City Harvest Apprenticeship and Business and Entrepreneurship for Local Foods graduate, said:

In the future, I'm really hoping to have my own incubator and grow produce and sell it within the community. I want to use my Windy City Harvest education to show communities that there is a better way to eat healthy and live healthy.

These words from Stacy, who was introduced to sustainable agriculture while incarcerated, echo the programme's ripple effect. At the same time that the Chicago Botanic Garden is helping to launch the careers of youth and adults with recognised barriers to employment, its programmes provide access to fresh, locally sourced food in low-income Chicago communities. Like other programme participants and graduates, Kimmons is now gainfully employed and contributing to Chicago's growing urban agriculture movement.

Case Study 8.2 *We Work towards a Sustainable Future: Jerusalem Botanical Gardens Social–Environmental Hub*

By Adi Bar-Yoseph, Jerusalem Botanical Gardens

The Jerusalem Botanical Gardens (JBG) established the Social–Environmental Hub to advance environmental awareness and encourage a sustainable lifestyle through support and promotion of social–environmental entrepreneurship. The Hub exemplifies the culmination of the gradual shift in the JBG's operating philosophy as it searched for its place in, and relevance to, the city in all its diverse complexity. The first step was to open the discourse and messaging so as to be more inviting. The second was to invite outside groups in for collaboration and co-creation of programmes and the third was the Hub – a platform for action and facilitator of an interdisciplinary professional network for change agents working towards sustainable community development.

As a university garden, the JBG knew well the challenges facing public gardens. The public perception was of a detached elitist institution into which one was invited only at the scientists' discretion, and this was reflected in the garden's landscape design, planting, programming and public relations. Strengthened by the significant social, political and security challenges characteristic of Israel and especially Jerusalem, this resulted in very low visitation and prompted the JBG to make a substantial shift in 2006 that resulted in a three step fundamental change – from openness to inclusion to interaction through dialogue.

The standard operating paradigm had essentially filtered out those who did not relate to the scientific, Hebrew language, message of the Gardens. To open it to a more diverse audience, the strong scientific messaging was slightly released and new programming and messaging were developed – traditional, 'community outreach'. This shift positively impacted visitation, quadrupling footfall and began the process which would fundamentally change the way the Gardens engages the public.

Under the new slogan 'plants grow people' the JBG also opened its gates to groups who wanted to create their own collaborative programmes within its grounds. These were primarily groups offering vocational rehabilitation, horticultural therapy or doing community work. Most are still operating in the Gardens today. As the number of groups and programmes at the JBG grew, a wider need and larger opportunity became apparent. As part of its mission to maintain biodiversity, it was imperative the JBG include within its operational philosophy engagement the complete arc of human diversity. Innovation appears where there is a discrepancy between need and what is on offer. In inviting new groups into the Gardens, whether for JBG run or externally run programmes, a dialogue emerged and gaps became evident. These were our opportunities for authentic engagement as the community told us what it wanted and where it would allow us, and even expected us, to step in.

As botanic gardens the world over re-think how they remain relevant, 'lower their fences' and engage their public, the JBG turned to these earliest social programmes and expanded the search beyond them with environmental professionals and activists from a wide range of disciplines – from gardening and agriculture to art to high-tech. In our 'market' or rather, community research, we discovered there was already a great deal being done in the city to promote sustainability and environmental awareness and the search for engagement was mutual. This was clearly a message for which there was an audience. How could we meet the city's inhabitants 'where they were' – engage them on their terms, grounded in their interests and without duplicating what already existed? What role can botanic gardens play in the efforts towards urban resilience and a sustainable future?

Botanic gardens are collections. In the past, that has meant privately holding something unique or exquisite. Today, collections are maintained to perpetuate their content for all (in itself a kind of sustainability). Sustainability is the preservation of resources – air, water, food, biological and cultural diversity. Plants and a deep appreciation of the importance of diversity are at the heart of this. Ironically, it seems there is a perceived dichotomy between plants in context of science and horticulture and plants as the basis for sustainable communities. Merging these perspectives is where public gardens have a responsibility to contribute their expertise and a way for them to meet a wider audience where their interests lie. At the JBG it is the Hub that does this, supplementing the botanical discourse with one which speaks to the most basic human needs and links them with plants, biodiversity and conservation.

This new paradigm for engagement brought new groups into the Gardens and created a space for them to convey their own content as well as to co-create content with the JBG or with other groups involved with the Hub. When interaction occurs on equal footing, it changes both parties; conveying our message within the context of sustainability created opportunities to venture into new realms such as new social, economic and environmental challenges, urban gardening techniques, urban planning, art, design, technology, even innovative social enterprise, thus addressing either directly or indirectly the full spectrum of human diversity, expanding visitation, public profile and the reach of the Gardens' traditional values and messages.

While the JBG could have continued to create programming and invite more groups in, it would have reached full capacity. Quickly. And so a new model was necessary. In its vision, goals and operational model the JBG Hub is based in familiarity with the community and is deeply rooted in the twenty-first-century tendency to open-sourcing and networking. To engage groups and individuals not only around issues of interest to them, but also through the mediums and modes they prefer, the Hub draws on the model of technological start-up hubs adapting it for the environmental realm to advance its goals and message by encouraging innovation and entrepreneurship. After dozens of meetings with activists, programme leaders, and public opinion shapers to map and understand the professional community on the city, the Hub focused its efforts on two overarching themes – professional development and networking. By creating a professional network for those working in the social–environmental field, the

JBG contributes to a crucial global cause closely linked with its core values and mission while creating a new platform on which these values can be conveyed and promoted. The Hub offers these groups professional assistance from the JBG experts supplemented by external support staff and consultants. More importantly, it enables and encourages collaboration and idea and resource sharing among members, matching organisations with complementary assets and needs to better affect change by advancing each group's specific goals as well as those of the community. Working in this model maximises existing capabilities as the community supplies its own needs with the Hub functioning mostly as facilitator, only directly supplying in its fields of expertise. To further facilitate this, a designated co-working space will be built for the Hub within the Gardens – a 'home' where these groups can work and interact informally among themselves and with the Gardens.

The interdisciplinary approach is fundamental. First, creating a sustainable future requires a deep behavioural shift which can only be achieved through solutions in all elements of life, second, this is the basis on which to attract new audiences and finally, the projects arising from interaction between professionals in different fields are more likely to be successful and infinitely more interesting. Three short examples show how wide-ranging this has become.

First and most closely aligned with the Gardens' traditional role is a local CSA farm working with teenagers who have dropped out of school. They built a hydroponic greenhouse on the JBG's undeveloped land and, alongside growing vegetables, they will run educational programming developed together with the JBG education department. Second, in the technological sphere, two groups bear mentioning; a green IT company developing smart gardening solutions will install its prototype in the Hub therapeutic greenhouse and a start-up working to revolutionise water management in Africa may take the hydroponic greenhouse off-grid. Technology developments hold much potential both in impact and in their potential to contribute to the sustainability of communities and the gardens themselves. Perhaps most innovative, is a group promoting a complementary local currency. Within its network the Hub uses this local currency as a sophisticated barter system to encourage interaction between its members. It also supports the venture as part of the commitment to a wide approach to sustainability. The benefit of complementary currencies is that by increasing trade within the community it lowers costs (monetary and environmental) incurred due to transport. Furthermore, it strengthens resilience of communities by increasing diversity and keeping currency within the community, semi-detaching buying power from standard income capacity. To date the JBG Social-Environmental Hub has mapped approximately 100 organisations striving for social-environmental change through urban agriculture, community work, art and design, technology, business and social business, most operating locally but some with international aspirations.

The JBG's operational paradigm shift led, in each of its stages, to an expansion in its audience. In taking a more open stance towards the city, footfall increased 400 per cent and this figure continues to rise. Inviting outside organisations to see the JBG as a resource for their own programming needs resulted in 20 groups operating within the grounds. The shift to equal interaction through dialogue led to the development of a network of 70 organisations that are in contact with the JBG and with whom its impact is multiplied exponentially.

Case Study 8.3 *We Advocate for the Richness of Diversity in Ecosystems and in Social Systems: Science Career Continuum*

By Amaris Alanis-Ribeiro, Kayri Havens, Andrea Kramer and Nyree Zerega, Chicago Botanic Garden and Northwestern University

The US has over 17,000 species of native vascular plants. Nearly one-third of them are considered threatened (NatureServe, 2012 a,b) and may require conservation action to prevent extinction. Preserving America's natural heritage will require that the large array of conservation agencies and organisations work collaboratively to provide much-needed plant conservation capacity. Most of the environmental grand challenges facing our planet require botanical expertise to solve. Whether it is food security, climate change, biodiversity conservation, managing invasive species or habitat restoration, botanical knowledge plays a role in finding solutions. But botanical capacity, which includes human resources, research funding and infrastructure and education programmes, is declining in the US and around the world (Kramer *et al.*, 2013). In the US, plant conservation is extremely under-resourced in comparison to animal conservation (Havens *et al.*, 2014). Additionally, there is a severe shortage of botanists in federal agencies with nearly half of them planning to retire by 2020. Because many of universities have dropped botany programmes, it is becoming more difficult to find botanically trained staff to refill these positions (Kramer *et al.*, 2010). The nation's science and land management agenda is suffering as a result.

The Chicago Botanic Garden's education and research departments are collaborating to help turn around this trend by providing a continuum of educational and research opportunities for students from middle school through graduate school that not only increases botanical capacity, but also aim to increase diversity in the fields of science, technology, engineering and mathematics (STEM), and especially science. Although nearly 30 per cent of the U.S. population is black or Latino, only 13 per cent of scientists come from those backgrounds (Landivar, 2013). Several research studies have indicated the importance of diversity (i.e. the importance for business bottom line, fostering a global perspective, different perspectives leading to innovation, etc.). Diversity is especially important for the field of science because the complexity of current environmental challenges requires the creativity and innovative thinking that can only come from the interaction of multiple perspectives and approaches, not just scientifically, but culturally as well.

Programme Structure

The Science Career Continuum begins with Science First and College First, serving up to 65 middle-school and high-school students of colour from Chicago Public Schools. A free summer science immersion programme – combined with paid internships and mentoring for high-school students – improves these students' overall academic performance and puts them on a path to college. Once in college, College First graduates can apply for a Research Experiences for Undergraduates (REU) internship in Plant Conservation or a teaching assistantship with the Science First programme. Armed with an undergraduate degree, College First graduates can apply for five- and ten-month Conservation Land Management (CLM) internships.[1]

Conservation Land Management interns receive a stipend to assist professional staff at the Bureau of Land Management, National Park Service, or US Forest Service in one of 13 western states. Finally, students seeking a graduate degree are eligible for scholarships to the Master's or PhD programmes offered jointly by the Garden and Northwestern University.

Figure 8.3 Conservation and Land Management interns carry out a quadrat study in the prairie. For the colour version, please refer to the plate section. In some formats this figure will only appear in black and white.

Middle and High School

Science First (SF) and College First (CF), with over two decades of programming, are the first steps in a successful model for a continuum of environmental science that provides students of colour with a clear and accessible pathway from middle school to college and science or STEM careers. They have a proven track record of success; among 51 alumni from 2005–2011 who responded to tracking surveys, 84 per cent completed post-secondary education with 61 per cent in STEM majors.

Figure 8.4 College First students collect macroinvertebrates in the lagoons at the Chicago Botanic Garden. For the colour version, please refer to the plate section. In some formats this figure will only appear in black and white.

Addressing Accessibility and Relevancy
Science First and College First provide programming for Chicago Public Schools (CPS) students who are black, Latino, or multiracial, first generation college students and/or students that qualify for free or reduced lunch. The programme targets students who are not necessarily at the top of their class or may have an undeveloped interest in science. To address barriers to participation, programmes are free, including all programme materials, lunch and bus transportation (students live 20–30 miles away from the Garden). Older students also receive a stipend – crucial for students who are often expected to supplement household income or fund their own education. This investment in addressing accessibility allows students to participate who may have been excluded from other enrichment programmes due to fees or GPA requirements.

Students engage with the Garden's 385-acre campus to foster social and cultural interest in science. Many have little experience interacting with nature due to life in heavily urbanised, often unsafe environments, so it is important to make cultural ties and illustrate the connections of a botanic garden with students' everyday lives. When building the curriculum, instructors reference students' applications, where students identify interest areas. As students learn about contemporary topics such as climate change, sustainable urban development and environmental justice, they reflect on how their communities are affected. In addition, participants choose their own research interest and design their final project/experiment; older students select their desired internship positions.

College and STEM Major Preparation and Persistence
Once juniors and seniors in high school, College First students have the opportunity to earn college credit in environmental science. Engaging in college-level curriculum puts students at an academic competitive advantage, but research demonstrates that academic preparedness is not enough to ensure college completion among low-income first generation students. To address this issue the programmes emphasise comfort with science equipment, development of twenty-first-century skills, and provide additional supports to encourage college enrolment and persistence.

The Chicago Botanic Garden provides an entirely new context to science unavailable in a typical CPS classroom. Students use the Garden as a 'living laboratory' and in the Garden's Plant Science Centre, students are trained to use advanced scientific instruments in nine highly specialised laboratories. Students use geographic information systems (GIS), scientific databases, specialised microscopes, an automated soil analyser, DNA sequencers and more. The hands-on internships prepare students for lab work in college, build STEM transferable skills, and exposes them to possible careers related to the environment. Given the rapidly changing and increasingly globalised STEM economy, it is equally important for students to develop twenty-first-century skills, including effective reasoning, use of systems thinking and problem solving. Because disseminating information is critical to the success of science and any environmental initiative, the SCC teaches students to communicate using a variety of platforms. Final presentations exemplify how skill development is fostered along a trajectory; communication skills expectations for final presentations are scaffolded from presenting one-on-one to presenting in a large auditorium.

For participating students, community around science and academia is a high priority. Students often do not have a personal relationship with a science professional and thus lack access to the informal networks that are critical to persisting in the field. The mentoring components of Science First and College First build an ecosystem network to foster the sense of belonging for students of colour. High school students work side-by-side with college undergraduates, graduate students and PhD scientists. This stair-step approach helps students understand and aspire to the next phase in the education continuum. The link between the Garden's informal education programmes and the plant science and conservation department creates a space for students to learn within a structured framework of professionals, including those in the Garden's joint graduate degree programme in Plant Biology and Conservation with Northwestern University. Mentorship builds community, crosses cultural divides, and engenders compassion. Students' mentorship of younger participants teaches them the value of collaborative work, while professionals improve their communication to lay audiences and gain a diverse perspective on subject matter.

College and Beyond

The Garden continues its support by offering opportunities for students in college and post-graduation. The Garden, with colleagues from partner institutions, hosts a 10-week summer Research Experiences for Undergraduates (REU) programme. This programme offers undergraduate participants an opportunity to explore a diverse array of scientific fields related to plant biology and conservation. Travel, room and board, and research costs are covered by the programme, and participants receive a $5,000 stipend.

Students work out of the Daniel F. and Ada L. Rice Plant Conservation Science Centre. They are trained in all stages of research, from hypothesis formulation through experimental design, data collection, analysis and ultimately presentation of results through a public symposium. There are often opportunities to participate in the professional community of practice by presenting at national scientific meetings or publishing findings in peer-reviewed journals. The REU interns interact closely with doctoral and Master's degree students and are encouraged to serve as research mentors for teens participating in the College First, as part of the step-wise mentoring system that allows younger students to envision themselves in the next phases of a science career.

Recent college graduates can continue their education and exploration of conservation science through the Conservation and Land Management Programme (CLM), a partnership between the Garden and federal land management agencies. Paid 5–10 month internships provide participants the opportunity to put their education and skills to work in a real-life setting. The majority work with partners in the western US, including the Bureau of Land Management, National Park Service, US Fish and Wildlife Service and US Forest Service. With over 1,000 interns placed since its inception in 2002, CLM provides partners with a valuable resource: young, knowledgeable, enthusiastic college graduates who are passionate about and seeking to gain hands-on experience in conservation and land management. Likewise, interns learn what it is like to work for a federal agency, explore their career goals, and expand their résumés. They gain experience with new landscapes, habitats and species diversity, which is valuable for both professional and personal growth. As one former intern stated, 'For future CLMers, I would like to say this: do not let this opportunity get away from you! This is an amazing programme and you will learn a lot!'

The final steps in the pathway to a research career in conservation science are the Master's and doctoral programmes in Plant Biology and Conservation offered collaboratively through the Chicago Botanic Garden and Northwestern University. These programmes provide students with advanced training in plant and soil ecology, evolution of plants and fungi and applied conservation theory and methods. Students take courses at both the Garden and Northwestern and interact with researchers and faculty from both institutions. The PhD programme aims to foster an academic and research environment that allows students to gain the experience, skills and knowledge required to become scholars, leaders or practitioners in plant biology and conservation.

Case Study 8.4 *We Help Build Local Capacity: Community-Based Floriculture Industry Development Project*

By Yves Nathan Mekembom, Limbe Botanic Garden

The Limbe Botanic Garden, Cameroon, was established by the German colonial government in 1890 as an agricultural research station with the purpose of researching plants of economic and medicinal interest, crops such as cacao, bananas, rubber and pineapple. Today, the Garden is a technical operational unit under the Ministry of Forestry and Wildlife and has continued that work in domestication of non-timber forest products and tropical cut flower cultivation. One such project is the community-based floriculture industry development project, which had as its goal to evaluate and upgrade Cameroon's efficiency in the production and marketing of high-quality tropical cut flowers that meet international horticultural standards. The Community-Based Floriculture Industry Development Project (CFIDP) aimed to understand the cost and scale at which it might be possible for Cameroon to develop a strong floriculture industry.

The project's key objectives include:

- To understand what tropical cut flowers and nursery plant species, packaging and marketing avenues are best for Cameroon.
- To improve understanding of production, disease prevention and pest control of tropical cut flowers.
- To identify potential plant species for development of unique flagship products to be introduced into the world's floriculture market.

With support from the US Department of Agriculture, the project began in 2003 by working with the Cameroonian community to introduce local farmers to the UN FAO Good Agricultural Practices (GAP) through training workshops and radio broadcasts. Farmers also worked with the project's organisers to select pilot farms on which GAP could be used, test cultivars and propagate good quality seeds and seedlings. Surveys of the local and European markets were conducted to understand whether a floriculture industry could be developed in Cameroon.

At the national and international levels, collaborators worked to create a market for these new products, analysing global floriculture market trends and identifying potential plant species for development of unique flagship products to be introduced into the world marketplace. At the local level, eroded and degraded land was converted into cut flower farms. Educational campaigns increased awareness of the importance of the native species in cultivation and encouraged their use as landscape plantings to further decrease erosion.

The creation of a global market and the local capacity to fulfil resulting demand required the investment of stakeholders from local farmers and businesses, to government ministries and scientific research. It required a collaborative approach, with attention to the needs of all.

Collaborators included the Ministry of Agriculture and Rural Development, Cameroon, Ministry of Town Planning, Cameroon, University of Buea, Cameroon, the Cameroon National Herbarium and Royal Botanic Gardens, Kew.

There were two primary challenges in this work. The first was getting initial buy-in at the local level, convincing local farmers and businesses that changing their crop production and taking time for GAP training would result in a successful new commercial industry. Likewise, creating the demand and assuring the global market that there would be a steady and reliable supply of materials was also a challenge. However, by bringing all the stakeholders together with their multiple areas of interest and leveraging their expertise, they were successful in both. A critical component of these accomplishments was the inclusion of indigenous populations and the incorporation of traditional ecological knowledge into conventional conservation efforts.

Since the project's inception, flower production has become one of the main income-generating activities in the communities around Limbe. The number of participating farmers has tripled since the beginning of the project, and the cultivation of these flowers is extending to other nearby localities, including Yaounde, Bamenda, Douala and Bafoussam. There is now a consistent, dependable flower supply, and in fact local flowers now dominate the local market. This has resulted in an improved local economy and transformed the region's landscape. The cultivation of native flower species has also resulted in successful *ex situ* conservation of biodiversity and a substantial reduction in erosion on the now farmed land. These species are now also being used for local landscaping, further decreasing erosion. Limbe Botanic Garden has also seen an increased number of visitors as a result of the project.

Acknowledgements

We thank the following for contributing the four case studies included in this chapter: Angela Mason (Chicago Botanic Garden), Adi Bar-Yoseph (Jerusalem Botanical Gardens), Amaris Alanis-Ribeiro, Kayri Havens, Andrea Kramer and Nyree Zerega (Chicago Botanic Garden and Northwestern University) and Yves Nathan Mekembom (Limbe Botanic Garden).

Note

1. See CLM Internship Programme blog 2013 http://clminternship.org/blog.

References

Havens, K., Kramer, A.T. and Guerrant, E. O. (2014). Getting plant conservation right (or not): the case of the United States. *International Journal of Plant Sciences*, 175: 3–10.

Kramer, A., Havens, K. and Zorn-Arnold, B. (2010). Assessing botanical capacity to address grand challenges in the United States. 64 pp. plus appendices. Available online at: www.bgci.org/files/UnitedStates/BCAP/bcap_report.pdf [accessed March 2017].

Kramer, A. T., Zorn-Arnold, B. and Havens, K. (2013). Applying lessons from the US Botanical Capacity Assessment Project to achieving the 2020 GSPC targets. *Annals of the Missouri Botanical Garden*, 99: 172–179.

Landivar, L. C. (2013). Disparities in STEM Employment by Sex, Race, and Hispanic Origin. American Community Survey Reports, ACS-24. US Department of Commerce, Economics and Statistics Administration, US census bureau. Available online at: www.census.gov/prod/2013pubs/acs-24.pdf [accessed March 2017].

NatureServe (2012a). NatureServe Explorer Online Database. Available online at: www.natureserve.org/explorer [accessed March 2017].

NatureServe (2012b). NatureServe Conservation Status. Available online at: http://www.natureserve.org/explorer/ranking.htm [accessed March 2017].

9 · *Botanic Gardens and Conservation Impact*
Options for Evaluation

SARA OLDFIELD AND VALERIE KAPOS

The loss of biodiversity is one of the major challenges faced by the world today. However, as we have seen in previous chapters, the importance of plant diversity and therefore the rationale for its conservation, have not been globally appreciated. Botanic gardens play a range of important roles in support of biodiversity conservation; but the scale of the plant conservation challenge is daunting, and the resources available for plant conservation are limited. There is, therefore, a very clear imperative for botanic gardens to make best use of limited resources for plant conservation and to highlight successful approaches and outcomes in order to justify their allocation of resources to conserving plant diversity. To increase the impact of botanic gardens and enhance their credibility as conservation organisations able to secure scarce resources, it is important to consider their general strengths and constraints, to identify ways to measure and evaluate their conservation actions and outcomes, and to reflect on how botanic gardens' conservation actions can be scaled up to meet global challenges.

9.1 Botanic Gardens: How They Operate and How They are Perceived

Botanic gardens are by their very nature complex multifunctional organisations with conservation being but one component of their work, though arguably the most important when considered in the context of the biodiversity crisis. As noted in Chapter 4, most botanic gardens share a number of characteristics that set them apart from other institutions and can have a particular relevance to conservation. Botanic gardens work at the interface between science, education and practice with a broad skills base. They are part of their own local and regional or in some cases national, communities and are able to influence decisions that affect conservation at different levels. Additionally, many botanic gardens have a degree of institutional stability, sometimes with a long history, and are able to take a long view: a perspective that is equally important in garden management and in conservation.

The extent to which botanic gardens build on these attributes to be actively involved in conservation varies considerably according to their individual missions, capacity and resources. The main purpose and roles of a botanic garden may be perceived in different ways by the staff, visiting public and society at large (see Box 9.1). Traditionally botanic

---

Box 9.1 *Perception of the Roles of Australian Botanic Gardens*

Botanic gardens are amongst the top four most visited cultural sites in Australia. A recent paper reported that 35 per cent of the Australian adult population visit a botanic garden at least annually – comparable with zoos and aquaria (37 per cent). The main purpose for visits to botanic gardens was recreation – 31 per cent visited to 'view plants', 19 per cent to 'walk or exercise', 12 per cent for a 'family outing' and 10 per cent to 'relax or read'. Only a small proportion (8 per cent) attended specifically to 'learn about plants' (with a further 4 per cent attending for a 'guided walk'). Twenty per cent of visitors were unaware that botanic gardens play a role in conservation. Whereas the main role of botanic gardens, as identified by visitors, is for 'recreation', volunteers at the gardens consider education to be the main focus, and the botanic gardens themselves, as organisations, consider conservation to be their main focus.

The majority of Australian botanic gardens are operated by local or state government and derive about 80 per cent of their income from government funding. Funding is increasingly dependent on visitor satisfaction measures as public sector models of governance in Australia are required to adopt more business-like managerial approaches. This may in part explain the perception of gardens as 'leisure centres' by the visiting public.

The authors of the paper conclude that, 'rather than considering these findings from the perspective of conflict, more is to be gained by considering the complementary role of the three groups' and the ways in which conservation, education and recreational activities can inform and enrich each other.

Source: Moskwaa and Crilley (2012)

gardens have not been perceived as conservation agencies and this may undermine their overall impact in addressing biodiversity conservation. The lack of recognition may partly result from the outdated perception that *in situ* conservation is the only valid, or at least the preferred option for saving species from extinction. The division between *in situ* and *ex situ* conservation is becoming increasingly blurred and may become irrelevant in a rapidly changing world (as discussed in relation to botanic gardens in Chapters 4 and 5 and more broadly by Pritchard *et al*. 2012).

Despite these variations in perception, there can be a tendency for botanic gardens (in common with other organisations such as zoos and aquaria) to justify their existence and funding base in terms of contributing to conservation, whereas in many cases this forms only a small part of their work. As noted by Fa *et al*. (2011), there is a declared intention by zoos to help address worldwide declines in biodiversity but 'there are still discrepancies between zoos' stated conservation goals and their actual performance'. Inevitably this is also the case in the world of botanic gardens but there are differences. Fa *et al*. (2011) point out that to contribute more effectively to restoring biological diversity, zoos need to resolve how commercial and conservation aims can be made compatible and also establish an operational model that allows them (jointly) clear measurable outputs and outcomes in biodiversity conservation. The botanic garden community has gone a considerable way to addressing the second of these issues through supporting the development and implementation of the GSPC, as outlined in Chapter 1, which provides an international framework for plant conservation in its broadest sense. Relating to this, botanic gardens have also been involved in the formulation of regional and national conservation strategies (see for example Box 9.2).

At an institutional level, botanic gardens that claim to be conserving plant diversity need to be able to show clearly that they are doing so; this is only partly about what activities they are doing and how well (measuring performance), and should also be principally about what the outcomes of those activities are. For botanic gardens to demonstrate clearly the value of any and all aspects of their work it is important to be explicit about setting objectives that reflect a clearly defined organisational mission (see Chapter 8) and to identify the links between their activities and the achievement of outcomes that will help to meet those objectives. Ideally it should be possible to aggregate project- and programme-specific outcomes relating to specific activities to show their contribution to higher level departmental and organisational objectives and overall

> **Box 9.2** *The North American Strategy for Plant Conservation*
>
> The 2016–2020 North American Botanic Garden Strategy for Plant Conservation (NABGSPC) is intended to provide a foundation for meeting ambitious and critical plant conservation targets throughout North America with targets developed for 2020. Developed and led by four network organisations in the US – the American Public Gardens Association, Botanic Gardens Conservation International, US (BGCI US), the Center for Plant Conservation (CPC), and the Plant Conservation Alliance Non-Federal Cooperators Committee (PCA-NFCC) – it is recognised that plant conservation goals can only be met through collective, coordinated and enthusiastic action. The overall shared objective is to sustain a healthy, biodiverse world where the essential benefits of plants that support sustainable livelihoods, health and well-being can continue to be enjoyed.
>
> There are over 1,000 botanic gardens and similar organisations throughout North America according to BGCI's GardenSearch database, all of which have unique capabilities and programmes that could contribute to plant conservation. Each of these gardens can craft its own response to the regional, national and global conservation challenges of the North American Strategy in the context of its own institutional mission. The NABGSPC directly supports and incorporates the goals, mission and vision of the 2011–2020 GSPC which is: to understand, conserve and use sustainably the world's immense wealth of plant diversity while promoting awareness and building the necessary capacities for its implementation.

mission. Crafting a vision and evaluating outcomes can present significant challenges, especially in addressing the full diversity of a botanic garden's activities, which may comprise, for example, estate management, public services, horticulture, education and science (see below).

The defining feature of a botanic garden is the presence of documented plant collections. The main purposes to which these collections are put are research, education and conservation. All of these can, of course, contribute directly or indirectly to a broader conservation mission. All botanic gardens also have a managed site where they are located which can be a hub for a range of activities involving the skills and expertise of their staff, partners and visitors. Utilising their collections, sites, facilities and expertise, botanic gardens are undertaking a wide range of activities to solve biodiversity-related problems beyond the garden and the gardening world. To be effective, explicit problem analyses and/or theories of change are required that speak to the problems in the wider world that gardens aim to contribute to solving.

It is helpful for gardens to identify both those conservation-relevant outcomes achieved by their activities and those of wider programmes that they are contributing to. Conservation impacts are rarely achieved through a single set of actions or by a single institution; gardens should be explicit about how and what they contribute. In a recent review of plant conservation in the US, the significant role of botanic gardens in supporting conservation activities is highlighted noting the need for botanic gardens to work strategically with all NGO, community, government and industry sectors to achieve common goals (Havens *et al.*, 2014). It is interesting to note that the 50 US botanic gardens with the largest budgets spend an average of 8.1 per cent on conservation and research, and devote 8 per cent of staff time to these activities (Sharrock *et al.* 2014).

9.2 Measuring the Performance of Botanic Gardens

Performance measures such as key performance indicators (KPIs) are important within organisations to measure and demonstrate how well important activities essential for the success of the organisation are being implemented. This is a fundamental part of assessing the likelihood that the activities will have an impact but true evaluation depends on assessing an explicit link between the activity being undertaken, its intended outcome and the situation it aims to address. In general, performance measures are used with the expectation that good outcomes will be achieved as a consequence but this assumption needs to be tested and reported on at appropriate intervals.

It can be challenging for botanic gardens to define their own measures of success. Visitor numbers by age and background and numbers of school visits at different stages in education are common measurements made by botanic gardens, useful in monitoring 'success' from year to year and for benchmarking against different institutions. However, assessing the educational impacts of those visits is more difficult. Developing credible measures for scientific, horticultural or conservation outcomes *that will be recognised as valuable by the outside world* (including funders) can be considerably more difficult.

In plant science, many botanic gardens rely on academic metrics such as peer-reviewed publications, citation indexes and numbers of publications in high 'impact factor' journals. These are appropriate for some of the scientific research carried out by botanic gardens but they do not capture much of the wider plant science that botanic gardens are uniquely qualified to carry out. For example, plant taxonomy and systematics – the documentation and description of plant diversity – is a fundamental branch of plant science that has traditionally been carried out in herbaria associated with botanic gardens. Although plant names, and their associated descriptions, appear in peer-reviewed publications (e.g. floras and journals), these publications do not appear on citation indexes or in high impact factor journals. This means that despite the fact that a plant name (and its author) will be cited every time that name is used subsequently, these citations are not measured and the author receives no credit. Plant identification is another scientific discipline and service that herbarium-based botanists are uniquely able to provide. This activity is essential for survey and inventory work for plant conservation, for the authentication of useful plants and many other purposes. However, few botanic gardens include plant identification amongst their key performance indicators (Paul Smith pers. comm.).

A range of performance measures for botanic gardens and related institutions involved in *ex situ* conservation are given in Box 9.3. Some

---

Box 9.3 *Performance Measures for* Ex Situ *Conservation Facilities*

The performance measures given here are derived from a web-based survey to establish a baseline assessment of botanic garden contributions to conservation relating to the GSPC. Development of the performance measures also took into account the conceptual framework outlined by Salafsky *et al.* (2002).

Site and species management
- Hectares owned in native habitats.
- Hectares secured in native habitats (e.g. through conservation agreements).
- Hectares managed in native habitats (e.g. weeds removed).
- Hectares restored in native habitats (e.g. native plants added).
- Species in *ex situ* conservation collections.
- Number of accessions per taxon (genetic diversity measure).

- Species in formal reintroduction programmes.
- Formal recovery plan contributions.
- Recovery team membership.
- Reserve plan contributions.
- Species recovered.
- Species conservation assessments (e.g. red listing).
- Formal agreements established with other conservation agencies (local, national or international).
- Steps taken to avoid invasive introductions.
- Applied research projects focused on high-priority species or habitats, or both.

Education/awareness
- Informal – interpretation (in garden).
- Informal – outreach (beyond garden).
- Formal – primary and secondary (K–12).
- Formal – university undergraduate.
- Formal – university graduate.
- Formal – certificate programme/continuing education/symposia.
- Communications – peer-reviewed general science publications.
- Communications – peer-reviewed biodiversity/conservation publications.
- Communications – lay publications.
- Communications – alternative (e.g. web pages).
- Demonstrating environmentally sound and sustainable practices.
- Displaying species and landscapes that promote the conservation work of the institution and collaborators and do not harm native species or habitats.

Changing incentives (internal and external)
- Contributing to sustainable use (e.g. garden retailing that supports sustainable development of economically disadvantaged communities; promotion of ecotourism).
- Using market pressure (e.g. certification programmes for nursery-propagated stock; boycotting over-harvested taxa).

Law and policy (internal and external)
- Institutional mission and role directly linked to established priorities for conservation, capacity building, and/or sustainable development (institutional decision making driven by conservation ethic; steps taken to conserve resources institutionally through reduced resource use, recycling, responsible use of chemicals, protection of water quality, etc.).

- Policy development (policies openly available on the Internet).
- Certification/licensing of facilities by federal authorities (e.g. US Fish and Wildlife Service) for endangered and threatened species work.
- Advocacy or lobbying for conservation issues.

Institutional investment
- Percentage of operated budget spent on conservation internally (i.e. on the actions listed above, not on science generally).
- Percentage of operating budget spent on extra-institutional conservation projects, including capacity building in high-priority areas (e.g. shared positions with other conservation organisations; functional links between garden exhibits/interpretation and landscape-level conservation).
- Professional development/training in conservation (for all staff, not just conservation staff).
- Number and training of conservation staff (full-time, part-time and students/interns).
- Number of conservation volunteers.
- Grant funding secured annually.

Based on Havens *et al.*, 2006

of the measures relate to activities carried out by the 'garden' whereas others such as institutional mission and institutional investment apply to the institution as a whole. Some of the measures, such as hectares secured and hectares restored, are also useful measures of conservation outcomes.

9.3 Measuring Conservation Success

Since Usher's pioneering book, *Wildlife Conservation Evaluation* (Usher, 1986), there has been a substantial body of literature on both the importance of evaluating the impact of conservation actions and how best to do this (see for example, Sutherland *et al.*, 2004; Bottrill *et al.*, 2011). Much of the discussion has centred around the difficulty of demonstrating what would have happened without the particular conservation action being carried out (Ferraro and Pattanayak, 2006). To address this challenge, greater emphasis is now frequently placed on establishing a well-supported baseline prior to commencing a conservation project, making careful 'before and after' observations, and improving project design in ways that make evaluation more robust. Despite the increased recognition of the need for evaluation, it is widely felt that there is still insufficient adoption by conservation practitioners. Curzon and Kontoleon (2016) explored why this might be through a Delphi Survey of conservation experts. They found little consensus between experts on what best practice in conservation evaluation should be but general consensus that the importance of evaluation is firmly accepted, with progress being made in carrying it out. It was noted that funding for conservation projects was such a limiting factor that funds were rarely available for rigorous evaluation to be undertaken.

The field of conservation evaluation is itself diverse, encompassing many different aims and objectives which adds to the difficulty of evaluating success. Furthermore, different stakeholders may have different perspectives on what has been successful and what has not (see for example, Roe *et al.*, 2013). The merit of developing an explicit Theory of Change during the design of projects has gained widespread recognition, especially during the last decade and is now used, to evaluate proposals for funding, for example by the UK Government's Darwin Initiative. The concept involves defining long-term goals and then a sequence of intermediate outcomes that must be achieved to deliver these goals. By asking what preconditions will be required for each outcome along the way and examining the assumptions that have been made a causal pathway can be established which will connect the activities undertaken as part of the project with its long-term goals. For a discussion of the application of Theory of Change in conservation biology see Morrison (2014).

A key step in using Theory of Change approaches is the identification of an appropriate long-term goal and defining what success in achieving it 'looks like'. Conservation success has been defined as 'increasing the persistence of native ecosystems, habitats, species and/or populations in the wild (without adverse effects on human well-being)' (Kapos *et al.*, 2010). Depending on the interpretation of 'persistence in the wild' applied (and these may be shifting as discussed in Chapter 4), this definition can serve as a good basis for clarifying the ultimate goals for botanic gardens' conservation-related work. The pathways of outcomes leading to this ultimate goal are as varied as the many different types of conservation activity that gardens engage in. Kapos *et al.* (2010; following Salafsky *et al.*, 2008) recognised seven broad categories of conservation activity (see Box 9.4); two that involve direct management of conservation

Box 9.4 *Seven Broad Categories of Conservation Action and their Relevance to Botanic Gardens*

The following broad categories were identified by Kapos *et al.* (2010) as part of a process of identifying steps in achieving conservation success and a process of evaluation. Relevance to botanic gardens is briefly summarised with examples from this book or as referenced.

Site management: Botanic gardens are involved in the management of botanically important sites in a variety of ways. According to GardenSearch, 458 botanic gardens manage natural areas on their own property. An example is Limahuli Garden and Preserve in Hawaii (part of the National Tropical Botanic Garden) – see Chapter 4. Botanic gardens are also involved in managing other sites in partnership with other agencies – see the Madagascar example in Chapter 4.

Species management: Botanic gardens are renowned for their role in *ex situ* conservation as outlined in Chapter 5. Over 1,100 botanic gardens provide records of the plants they hold in *ex situ* collections to BGCI with 29 per cent of globally threatened species recorded (Sharrock *et al.*, 2014). In the US, members of the Center for Plant Conservation have monitored over 2,000 sites of threatened species and engaged in over 200 reintroduction programmes (Kramer *et al.*, 2013).

Research: Botanic gardens are involved in a wide range of research from traditional plant systematics and floristics to applied conservation research. They are developing new techniques for seed banking, plant propagation and reintroduction together with habitat management and restoration (Havens *et al.*, 2014). Botanic gardens are actively involved in assessing the conservation status of plant species – see Chapter 3 – using the information to develop recovery and action plans and identify priority sites for conservation.

Education: Education is one of the key roles of botanic gardens. A survey on the implementation of the GSPC by botanic gardens, carried out in 2010, indicated that Target 14 on education and public awareness is the most frequently implemented target. In the US it has been noted that 152 gardens have an education programme, with 110 providing education programmes for visitors and 45 providing education programmes at university level. Over 28,000 volunteers are engaged in education and outreach (Kramer *et al.*, 2013).

Capacity building: Botanic gardens are involved in building botanical and horticultural capacity and in the larger institutions capacity for practical conservation. Missouri Botanical Garden (MBG), for example, notes that capacity building underpins much of its work in support of plant conservation: 'Capacity building efforts are designed to benefit all citizens living in communities, helping to develop the knowledge, skills, and motivation to care for the environment and use resources sustainably.'

Policy: Botanic gardens influence biodiversity conservation policy at a variety of levels. The development of the GSPC was largely led by the botanic garden community as outlined in Chapter 1. National GSPC focal points are often individuals working for botanic gardens. Botanic gardens are also involved in CITES implementation.

Livelihoods: Botanic gardens are increasingly addressing livelihood actions. For example the Millennium Seed Bank Partnership (MSBP) of RBG, Kew, works with rural communities in Botswana, Kenya, Mali, Mexico and South Africa through a participative approach for selection, prioritisation and propagation of economically important indigenous species, supported by scientific research (Ulian *et al.*, 2016). Another example is the work of RBG, Jordan (see Chapter 6).

targets (species or sites), and five that influence conservation status indirectly, through work on policy and legislation, livelihoods, capacity building, conservation and research. As described throughout this book and briefly summarised in Box 9.4, botanic gardens are engaged in all seven types of conservation action.

For each of these categories of conservation action, working groups of conservation practitioners have developed generic conceptual models of the likely relationships between successful implementation, conservation-relevant outcomes and ultimately conservation impact, making explicit the linkages that are often assumed (Kapos *et al.*, 2010). These provide a useful framework for developing approaches and measures for evaluating conservation by botanic gardens and others.

Within botanic gardens, evaluating conservation success can be at institutional or project level. Some of the performance measures in Box 9.3 are useful starting points to consider in measuring activities and in some cases outcomes. Linking through to outcomes and long-term effects can, however, be challenging for *ex situ* conservation which is often about keeping options open for the future. Evaluation of the collections themselves can be relatively straightforward as discussed in Chapter 5, but evaluating their potential contribution to restoring species and ecosystems in the wild is more difficult. Attempts to do this looking, for example, at the genetics of plant species in the wild and in collections are described by Cires *et al.* (2013) and Cibrian-Jaramillo *et al.* (2013). In terms of planning collections, as botanic gardens share information on the holdings in their living collections, through the PlantSearch database, it is increasingly possible to be more strategic in collection development. PlantSearch and other databases can be used to analyse and address gaps in the *ex situ* holdings of the botanic garden community.

9.4 Measuring Botanic Garden Progress Towards the Targets of the GSPC

As outlined in Chapter 1, the GSPC is an internationally agreed strategy for the conservation of plant diversity agreed by Governments worldwide. Updated in 2012, the rationale of the GSPC notes the urgent concern that:

> ... many plant species, communities, and their ecological interactions, including the many relationships between plant species and human communities and cultures, are in danger of extinction, threatened by such human-induced factors as, inter alia, climate change, habitat loss and transformation, over-exploitation, alien invasive species, pollution, clearing for agriculture and other development. *(CBD, 2010)*

The ultimate goal of the GSPC is to halt the continuing loss of plant diversity. Clearly botanic gardens cannot achieve this alone but many, individually or collectively, are contributing directly to the 16 GSPC targets as has been demonstrated throughout this book. There is considerable merit in aggregating the combined achievements of the botanic garden community in alignment with the GSPC and showing how these activities relate to overall progress towards the GSPC targets both nationally and internationally.

Botanic gardens have used the GSPC targets in different ways. Some major gardens, for example Missouri Botanical Garden, have used the GSPC to align their institutional objectives and targets. Others have used the GSPC to think more about their role in conservation, for example, botanic gardens in New Zealand (Myers and Elliot, 2013) and others to guide plant conservation policy. For example the Royal Botanic Garden,

Tasmania, has reviewed its work against the targets, as has the Royal Botanic Garden Edinburgh (Blackmore *et al.*, 2011). Other gardens have taken responsibility for specific targets within a national context.

Another important outcome of the GSPC has been to stimulate discussion between botanic gardens and other conservation agencies at a national level. This can be seen, for example, in China. China's national response to the GSPC, China's Strategy for Plant Conservation, resulted from a unique interdisciplinary effort involving botanic gardens working under the umbrella of the Chinese Academy of Sciences together with the State Forestry Administration and the State Environmental Protection Agency. Bringing together different high level agencies increased awareness of the range of plant conservation options available. A review of the Chinese Strategy highlights the need for joined-up action to develop and strengthen the linkages at operational level (Gratzfeld and Wen, 2012).

A review of implementation of the first phase of the GSPC (2002–2010) by botanic gardens was undertaken in 2010 through a questionnaire survey of over 600 botanic gardens (Williams, *et al.* 2012). The study investigated the influence of the GSPC on botanic gardens, defining 'influence' as 'a change in the activities of the botanic garden'. The study also investigated the aspects of the GSPC that are being more commonly implemented by botanic gardens and considered how to promote the GSPC to target institutions to increase implementation. It was found that gardens that were networked globally and gardens with larger budgets were implementing more targets. Targets implemented tended to be aligned with existing institutional missions and aims. The survey found that Targets 1, 8, 14, 15 and 16 were well supported, with over 50 per cent of gardens identifying actions contributing to implementation of these Targets. Activities supporting Target 4 (*in situ* conservation), and Targets 6, 9 and 12 (relating to sustainable production of land, crops and other plants with livelihood importance) were less frequently reported with fewer than 15 per cent of responding gardens conducting activities supporting these Targets. The fact that conservation of socio-economic species and sustainable use of plants are the least implemented of the GSPC targets by botanic gardens is also noted by Paton and Lughadha (2011) who suggest that this is because these three targets are not considered traditional activities of botanic gardens.

Global progress in implementation of the GSPC was reviewed again in 2014 taking into account the activities of botanic gardens and other sectors (Sharrock *et al.* 2014). This review concluded that the GSPC had galvanised action in plant conservation and that significant progress was being made towards certain targets. It noted that botanical information was needed as a matter of urgency to address GSPC Targets 1, 2 and 5. These are effectively research and documentation targets without which progress in *in situ* conservation and sustainable use of biodiversity at the landscape level as required by GSPC Targets 4, 5, 6, 7 and 10 cannot be measured internationally. A consortium of botanic gardens and other scientific institutions is leading work on Target 1 which underpins the implementation of plant conservation as a whole. A similar approach may be needed to scale up action on GSPC Target 2. The GSPC Target 5 utilises information on threatened plant species as one means of defining Important Plant Areas, a task that some botanic gardens are contributing to in partnership with the NGO Plantlife International.

More recently a review paper for the Global Partnership for Plant Conservation (GPPC) considers that the GSPC targets have proved very successful in stimulating individual institutional action for plant conservation, including by botanic gardens, but that this has rarely been scaled up to the national, policy level (Sharrock and Wyse-Jackson, 2016). Whereas some countries including Brazil, China, Mexico and South Africa have developed national GSPC responses, most parties to the CBD have considered plant conservation within the context of their National Biodiversity Strategies and Action Plans. In general, specific plant conservation activities continue to be poorly integrated into national biodiversity policies, and plant conservation activities that contribute towards GSPC targets are often not fully incorporated into national biodiversity reports. Countries rarely connect the importance of plant conservation to national economic development priorities or even to overall environmental protection efforts.

The review paper notes that plants are still neglected in the broader biodiversity and sustainability debate. Lack of comprehensive data on the conservation status of plants (see Chapter 3) compared to other components of biodiversity (mammals, birds, amphibians, etc.) means that plants are rarely used as indicators of the status of biodiversity and policymakers and the public generally remain blind to the fate of plants.

Overall although a diverse range of excellent plant conservation tools, methodologies, approaches, procedures and initiatives, have been developed during the lifetime of the GSPC, efforts to apply them at national and local levels have been patchy at best (Sharrock and Wyse-Jackson, 2016). It is to be hoped that the marginalisation of plant conservation by governments and international agencies will be seen as a major opportunity and challenge to botanic gardens, stimulating redoubled effort. Achievement of GSPC targets on education and awareness of plants and their value to humanity needs to be scaled up in raising the profile and support for urgent plant conservation globally. This is a key role for botanic gardens, their networks and the communities of which they form part. Increasingly botanic gardens are looking at ways to grow their social role, working with local communities on common issues of social and environmental importance (Vergou and Willison, 2016). Finally, when it comes to evaluating their wider contribution, botanic gardens should consider reviewing whether or not their efforts will contribute to the achievement of the 2030 Sustainable Development Goals. These Goals ambitions that require the conservation of plant diversity but go far beyond this.

References

Blackmore, S., Gibby, M. and Rae, D. (2011). Strengthening the scientific contribution of botanic gardens to the second phase of the Global Strategy for Plant Conservation. *Botanical Journal of the Linnaean Society*, 166: 267–281.

Bottrill, M. C., Hockings, M. and Possingham, H. P. (2011). In pursuit of knowledge: addressing barriers to effective conservation evaluation. *Ecology and Society*, 16: 14.

Curzon, H. F. and Kontoleon, A. (2016). From ignorance to evidence? The use of programme evaluation in conservation: evidence from a Delphi survey of conservation experts. *Journal of Environmental Management*, 180: 466–475.

Cibrian-Jaramillo, A., Hird, A., Nora, O., Ma, H., Meerow, A. W., Francisco-Ortega, J. and Griffith, P. (2013). What is the conservation value of a plant in a botanic garden? Using indicators to improve management of *ex situ* collections. *The Botanical Review*, 79: 559–577.

Cires, E., Smet, Y., Cuesta, C., Goetghebeur, P., Sharrock, S., Gibbs, D., Oldfield, S., Kramer, A. and Samain, M.-S. (2013). Gap analyses to support *ex situ* conservation of genetic diversity in Magnolia, a flagship group. *Biodiversity and Conservation*, 22: 567–590.

Convention on Biological Diversity (CBD). (2010). Consolidated update of the global strategy for plant conservation 2011–2020. Available online at www.cbd.int/gspc/strategy.shtml [accessed March 2017].

Fa, J. E., Funk, S. M. and O'Connell, D. (2011). *Zoo Conservation Biology*. Cambridge: Cambridge University Press.

Ferraro, P. J. and Pattanayak, S. K. (2006). Money for nothing? A call for empirical evaluation of biodiversity conservation investments. *PLoS Biology*, 4, e105. doi: http://dx.doi.org/10.1371/journal.pbio.0040105.

Gratzfeld, J. and Wen, X. (2012). *China's Strategy for Plant Conservation (SSPC): Progress of Implementation with Special Emphasis on CSPC Target 8 and Interrelated Targets*. Richmond, UK: Botanic Gardens Conservation International.

Havens, K., Vitt, P., Maunder, M., Guerrant, E. O. and Dixon, K. (2006). *Ex situ* plant conservation and beyond. *BioScience*, 56(6): 525–531.

Havens, K., Kramer, A. T. and Guerrant, E. O., Jr., (2014). Getting plant conservation right (or not): the case of the United States. *International Journal of Plant Sciences*, 175(1): 3–10.

Kapos, K., Manica, A., Aveling, R. *et al.* (2010). Defining and Measuring Success in Conservation. In: Leader-Williams, N., Adams, W. M. and Smith, R. J. (Eds), *Trade-Offs in Conservation: Deciding What to Save*. Oxford: Blackwell Publishing.

Kramer, A. T., Zorn-Arnold, B. and Havens, K. (2013). Applying lessons from the US Botanical Capacity Assessment Project to achieve 2020 Global Strategy for Plant Conservation Targets. *Annals of the Missouri Botanical Garden*, 99: 172–179.

Morrison, S. A. (2014). A framework for conservation in a human-dominated world. *Conservation Biology*, 3: 960–964.

Moskwaa, E. C. and Crilley, G. (2012). Recreation, education, conservation: the multiple roles of botanic gardens in Australia. *Annals of Leisure Research*, 15(4): 404–421.

Myers, T. and Elliot, R. (2013). New Zealand botanic gardens and the Global Strategy for Plant Conservation. Proceedings of the 5th Global Botanic Gardens Congress. Available online at: www.bgci.org/files/Dunedin2013/Proceedings/Success%20Globally/Myers%20and%20Elliot%20New%20Zealand%20botanic%20gardens.pdf [accessed March 2017].

Paton, A. and Nic Lughadha, E. (2011). The irresistible target meets the unachievable objective: what have 8 years of GSPC implementation taught us about target setting and achievable objectives? *Botanical Journal of the Linnean Society*, 166: 250–260.

Pritchard, D., Fa, J. E., Oldfield, S. and Harrop, S. R. (2012). Bring the captive closer to the wild: redefining the role of *ex situ* conservation. *Oryx*, 46(1): 18–23.

Roe, D., Grieg-Gran, M., Mohammed, E. Y. (2013). Assessing the Social Impacts of Conservation Policies: Rigour versus Practicality. IIED Briefing Papers. International Institute for Environment and Development, London. Available online at: http://pubs.iied.org/pdfs/17172IIED.pdf. [accessed 11 April 2016].

Salafsky, N., Margoluis, R., Redford, K. H., Robinson, J. G. (2002). Improving the practice of conservation: a conceptual framework and research agenda for conservation science. *Conservation Biology*, 16: 1469–1479.

Salafsky, N., Salzar, D., Stattersfield, A. J. *et al.* (2008). A standard lexicon for biodiversity conservation: unified classifications of threats and actions. *Conservation Biology*, 22(4): 897–911, doi: 10.1111/j.1523–1739.200800937.x.

Sharrock, S. and Wyse-Jackson, P. (2016). Plant conservation and the Sustainable Development Goals. A discussion paper for the Global Partnership for Plant Conservation Conference, St Louis, US, 28–30 June. Working draft.

Sharrock, S., Oldfield, S. and Wilson, O. (2014). *Plant Conservation Report 2014: A Review of Progress in Implementation of the Global Strategy for Plant Conservation 2011–2020*. Richmond, UK: Secretariat of the Convention on Biological

Diversity, Montréal, Canada and Botanic Gardens Conservation International. Technical Series No. 81.

Sutherland, W. J., Pullin, A. S., Dolman, P. M., Knight, T. M. (2004). The need for evidence-based conservation. *Trends in Ecology and Evolution*, 19, 305–308. http://dx.doi.org/10.1016/j.tree.2004.03.018.

Ulian, T., Sacande, M., Hudson, A. and Mattana, E. (2016). Plant Conservation for the Benefit of Local Communities: The Useful Plants Project: In: Rakotoarisoa, N. R., Blackmore, S. and Riera, B. (Eds), *Proceedings of the UNESCO International Conference*, 'Botanists of the twenty-first century: roles, challenges and opportunities' held in September, Paris, France.

Usher, M. B. (Ed.) (1986). *Wildlife Conservation Evaluation*. London and New York: Chapman and Hall.

Vergou, A. and Willison, J. (2016). Relating social inclusion and environmental issues in botanic gardens. *Environmental Education Research*, 22(1): 21–42.

Williams, S. J., Jones, J. P. G., Clubbe, C., Sharrock, S. and Gibbons, J. M. (2012). Why are some biodiversity policies implemented and others ignored? Lessons from the uptake of the global strategy for plant conservation by botanic gardens. *Biodiversity and Conservation*, 21: 175–187.

10 · *Conclusions*

STEPHEN BLACKMORE, SARA OLDFIELD
AND PAUL SMITH

It should be clear from the contents of this book that plant conservation is flourishing in botanic gardens, both as a focus for research and in its practical implementation whether through *in situ* or *ex situ* programmes. However, in such an enormous field, with so many active practitioners, it is only possible to provide an overview of the current state of the field and to point towards the growing body of literature on the subject. Several recent reports stand out as excellent summaries of the state of play in plant conservation, notably those compiled by Sharrock *et al.* (2014) and GPPC (2016). These reports draw widely on the experiences of the international community of botanic gardens and show how their work contributes to the wider conservation agenda. Another recent report provides an overview of the state of the world's plants with important messages on plant conservation (RBG Kew, 2016).

As the scope of this book is so wide, we consider it helpful to draw out and reflect on key themes based on the issues discussed in each chapter. We do this by way of conclusion, adding ideas for scaling up the work of botanic gardens in addressing some of the major global challenges in a coordinated manner through the development of a Global System for Botanic Gardens.

We also emphasise, in concluding, the profound importance of the 2030 Sustainable Development Goals (UNDP, 2015) in shaping and directing the global community of botanic gardens in their strategic thinking and future plans. Of most direct relevance to both the traditional and future of work of botanic gardens is Sustainable Development Goal 15, which aims to sustainably manage forests, combat desertification, halt and reverse land degradation and halt biodiversity loss. These aims are in close accord with those of the Global Strategy for Plant Conservation that has guided the work of botanic gardens since 2002. The 12 targets defined for Sustainable Development Goal 15 connect with many of the recurrent strands of discussion in this book, and are shown in Box 10.1.

Box 10.1 *Targets of Sustainable Development Goal 15*

1. By 2020, ensure the conservation, restoration and sustainable use of terrestrial and inland freshwater ecosystems and their services, in particular forests, wetlands, mountains and drylands, in line with obligations under international agreements.
2. By 2020, promote the implementation of sustainable management of all types of forests, halt deforestation, restore degraded forests and substantially increase afforestation and reforestation globally.
3. By 2030, combat desertification, restore degraded land and soil, including land affected by desertification, drought and floods, and strive to achieve a land degradation-neutral world.
4. By 2030, ensure the conservation of mountain ecosystems, including their biodiversity, in order to enhance their capacity to provide benefits that are essential for sustainable development.
5. Take urgent and significant action to reduce the degradation of natural habitats, halt the loss of biodiversity and, by 2020, protect and prevent the extinction of threatened species.
6. Promote fair and equitable sharing of the benefits arising from the utilisation of genetic resources and promote appropriate access to such resources, as internationally agreed.
7. Take urgent action to end poaching and trafficking of protected species of flora and fauna and address both demand and supply of illegal wildlife products.
8. By 2020, introduce measures to prevent the introduction and significantly reduce the impact of invasive alien species on land and water ecosystems and control or eradicate the priority species.
9. By 2020, integrate ecosystem and biodiversity values into national and local planning, development processes, poverty reduction strategies and accounts.
10. Mobilise and significantly increase financial resources from all sources to conserve and sustainably use biodiversity and ecosystems.
11. Mobilise significant resources from all sources and at all levels to finance sustainable forest management and provide adequate incentives to developing countries to advance such management, including for conservation and reforestation.
12. Enhance global support for efforts to combat poaching and trafficking of protected species, including by increasing the capacity of local communities to pursue sustainable livelihood opportunities.

It is important to note that other SDGs are also directly relevant to the work of botanic gardens (Blackmore, in press), in particular:

Goal 2: End hunger, achieve food security and improved nutrition and promote sustainable agriculture.
Goal 4: Ensure inclusive and equitable quality education and promote lifelong learning opportunities for all.
Goal 11: Make cities and human settlements inclusive, safe, resilient and sustainable.
Goal 13: Take urgent action to combat climate change and its impacts.

As the Sustainable Development Goals becoming increasingly integrated into policy at the international level and in countries around the world, they will provide important opportunities for botanic gardens to engage with the many other implementing agencies. In so doing they will be able to deliver on their unique potential to be the conduit for understanding and conserving the full spectrum of plant diversity.

10.1 Key Themes from Chapters

Chapter 1 emphasises that succeeding in conserving plant biodiversity is a fundamental prerequisite to securing the best possible future for humanity and for the enormous diversity of species with which we share this planet. So many other endeavours, from securing adequate supplies of food and fresh water to moderating the impacts of climate change depend upon the continuing existence of species-rich vegetation suited to the prevailing conditions of ecosystems around the world. This far from trivial challenge of mounting a fundamental defence of the plant kingdom will require botanic gardens to maximise their contribution and to work directly in partnership with many other players at the practical and policy level.

Chapter 2 focuses on research using molecular biology aimed at completing the documentation of plant diversity, a task which remains far from complete, even as undiscovered or yet to be described plant species are becoming extinct around the world (Pimm and Joppa, 2015). We cannot afford to see plant taxonomy and nomenclature as old fashioned science or indulgent academic pursuits – we need to know how many plant species there are and where they occur. One of the more successful outcomes of the GPSC has been to collate information for *The Plant List* as referred to throughout this book and to initiate the World Flora Online. New technologies enable greater understanding of plant diversity and have many applications in plant conservation (as Corlett (2016) has also emphasised). One important example, is the potential to identify the species of fragments of plants, from mixtures in medicines to sawn logs of timber, which can underpin the urgent action called for to end poaching and trafficking of protected species, and address both demand and supply of illegal wildlife products as called for in Target 7 of SDG 15. Whilst advanced molecular biology facilities may at present be beyond the reach of many botanic gardens, especially those in developing countries, falling prices and the advent of mobile, field-based laboratory equipment are set to expand access to them.

Chapter 3 explores the major threats to plant diversity, noting that habitat loss continues to be the greatest driver of biodiversity loss. Overexploitation of plant resources whether for immediate, local use of for legal or illegal trade is one of the greatest threats to plant diversity and, therefore, challenges to conservation. The point is made that even when there is a long established tradition of using a particular plant product and where that activity has been sustained for centuries, a tipping point will be reached if the scale of exploitation increases. This happens, for example, when once locally used and traded plant products become globally traded and, in today's world with a human population of 7.4 billion, a steadily growing market exists for many such species. The result is that new markets open up for the relatives of species already driven to extinction, or threatened to the extent that they can no longer by harvested from the wild. Invasive species, themselves often dispersed via global trade routes, pose a growing threat. Botanic gardens have recently started to embrace this challenge, responding through the establishment of plant sentinel networks (Barham *et al.*, 2015) to prioritise early detection of invasive species. The additional threats posed by climate change are exacerbating the precarious situation for many plant species in the wild, for example, island endemics and species with limited ability to adapt. At the time that ecological resilience through species diversity is needed to help buffer the impact of climate change, plants are increasingly under threat. It is unfortunate that the pace of evaluating the conservation status of plants for the IUCN Red List has been so slow and that plants are not adequately represented in biodiversity policy indicators as a consequence.

Chapter 4 draws upon the experiences of a tropical botanic garden situated in a biodiversity hotspot impacted by habitat loss and development to provide a perspective on the contribution of botanic gardens to *in situ* conservation of plants. The authors argue for an expanded definition of *in situ* conservation which includes, but goes far beyond, the tradition of conserving threatened species in protected areas. The case is made for conservation even in urban areas, citing the growing emphasis placed on urban forests and the objectives of Sustainable Development Goal 11 with its concern for cities and human settlements (see, for example, Elmqvist *et al.*, 2013; Pinto *et al.*, 2016). The need for a more complete taxonomic inventory discussed in Chapter 2, especially in the tropics, is highlighted. The importance of establishing new networks and partnerships, a recurrent theme throughout the book, is forcefully made.

The collections held in botanic gardens, arboreta and seed banks and their roles in research and in *ex situ* conservation are the focus of Chapter 5. The importance of seed banks and the protocols they adopt, lead on into a detailed discussion of the exceptional species which are not suited to conventional seed banking and the alternative modes of preservation that are being developed. A clear statement is made of the need for future efforts to address the information gaps, research the biology of exceptional species and to address the challenges of funding, communication and coordination. The case is made that the niche of botanic gardens in *ex situ* conservation needs to move beyond the collection of plants as an insurance policy to protect against the loss of wild diversity and to become a major source of appropriate plant material to be used in *in situ* conservation, including in ecological restoration. However, the exchange of plant materials is covered by international agreements on access and benefit sharing and as this chapter makes clear there are problems to be overcome before the exchange of non-crop plant material becomes as straightforward and effective as it needs to be if efforts to restore degraded ecosystems are to move forward as they urgently need to.

The role of botanic gardens in restoration, as discussed in Chapter 6, ranges from species recovery programmes involving the reintroduction of rare plants into the wild through to community and landscape restoration. Restoration ecology of plants is introduced as a lively field of research encompassing disciplines as diverse as seed biology, soil ecology, pollination biology, community ecology, and population genetics. Six major obstacles are described which must be overcome if large-scale ecological restoration is to succeed. The first of these is the fact that, around the world, the rate limiting step for restoration projects is the difficulty of obtaining substantial quantities of genetically appropriate seed. In too many situations, fast growing, and often non-native species are used to restore degraded vegetation, either through lack of understanding or simply because such species are readily available. Clearly one of the greatest challenges to botanic gardens is to make available a wider range of species for restoration, with appropriate provenance, if they are to effectively use their living collections and seed banks as sources of material for reintroduction and restoration. Generally botanic gardens hold collections of plant species that are unlikely to be readily available anywhere else. The necessity for long-term monitoring with the potential for citizen science projects is stressed, as is the need to 'balance the needs of the plant community with the needs of the human community'. The interwoven challenges of climate change, assisted migration and maintaining the genetic diversity of material bulked up from seed collections is considered. The chapter highlights the importance of the Ecological Restoration Alliance of Botanic Gardens (Aronson, 2014) as a new initiative that aims to share experiences and to develop practical solutions to the challenges referred to previously.

Chapter 7 opens up the contribution of botanic gardens to global food security and global fuel security, the issues addressed by SDG 2: End hunger, achieve food security and improved nutrition and promote sustainable agriculture; and SDG 7: Ensure access to affordable, reliable, sustainable and modern energy for all. The point is made that although botanic gardens played an important historical role in the introduction of many important plantation crops, from tea to rubber, they are, in general, no longer perceived as relevant. In fact, however, contemporary botanic gardens can contribute to the food security agenda by using their living collections in trait-based crop research, by conserving landraces and crop wild relatives and by engaging with their visitors to raise awareness of the issue. The authors argue that whilst botanic gardens are uniquely placed to carry out such research they tend not to realise the opportunity to contribute. Two case studies show the enormous potential impact of botanic garden research on the engineering of C_4 photosynthesis and on introducing nitrogen-fixing root nodulation into cereal crops. The importance of plant-based biofuels, including bioethanol and biodiesel are used to highlight the potential role of living collections in botanic gardens, taking the biofuel garden at Xishuangbanna Tropical Botanical Garden as an example. A further case study, on the development of a collection of microalgae for research on algal biofuels at the Cambridge Algal Innovation Centre, challenges botanic gardens to think more broadly about the collections they can develop. Finally, Chapter 7 introduces the concept and impact of 'plant blindness' (Wandersee and Schussler, 1999), setting out the very significant opportunity that botanic gardens have to engage with their visitors and wider stakeholder communities.

Chapter 8 develops and enlarges on the theme of botanic gardens engaging with the broadest and most diverse audiences. It describes how Chicago Botanic Garden has radically redefined the way it thinks about both itself and about its audiences. The authors advocate the profound importance of celebrating human diversity and engaging authentically with those they wish to connect with. In 2014, Chicago Botanic Garden adopted an inspiring new mission statement: 'We cultivate the power of plants to sustain and enrich life.' Contained within this statement is the idea that 'we cultivate' refers not just to plants but to the minds of everyone, whether staff, students or ex-offenders, enrolled in the highly innovative Windy City Harvest, urban agriculture programme. The argument is made that botanic garden leaders must become agents of change if botanic gardens are to achieve their full potential in this era of global change.

Chapter 9 focuses on the need for botanic gardens to be able to demonstrate how they are delivering upon their missions. A recurrent theme in several chapters has been the gap between how botanic gardens are perceived by wider society and how they think about themselves. It is important, therefore, that the strategies of individual gardens and the broader strategies that they individually and collectively contribute to (such as the Global Strategy for Plant Conservation) are understood and gardens are able to show that they are delivering. Developing appropriate measures of performance has been an emerging topic in the botanic garden community in recent years and a parallel discussion in the wider conservation community has focused on how to evaluate the success of conservation programmes (see for example, Kapos et al., 2008, 2009). The authors argue persuasively that, for botanic gardens to overcome their marginalisation in plant conservation they need a better framework for evaluation to sit alongside the strategic frameworks and action plans that have been developed. Beyond this, botanic gardens need to ensure that their conservation efforts become more integrated and effective. How they might do this is explored next.

10.2 Botanic Gardens and the Future of Plant Conservation

The major conclusion emerging from the content of this book and from recent conference discussions and debate is the need for a cost-effective, rational Global System for Plant Conservation.

The concept of such a Global System has been developed and described by Smith (2016) in comparison with the endeavours of the global crop research community. Despite its importance to food security, much of the world's crop diversity is neither safely conserved, nor readily available to scientists and farmers who rely on it to safeguard agricultural productivity. Crop diversity is being lost, and with it the biological basis of our food supply. Given the urgent need to achieve food security in the face of a changing climate and burgeoning human population, the crop research community has developed the concept of a cost-effective, rational Global System for the conservation and sustainable use of plant genetic resources in food and agriculture (FAO, 2011). This Global System, established by the Food and Agriculture Organization of the United Nations (FAO), comprises elements of policy, planning, a review process, physical infrastructures, human resources, germplasm collections and data. It consists of:

- The International Treaty on Plant Genetic Resources for Food and Agriculture (PGRFA).
- The global plan of action for PGRFA.
- A review process (state of the world's PGRFA).
- A network of international institutions and crop collections.
- A global portal of accession-level data (Genesys).
- A universal gene bank information management system (GRIN Global).
- Advanced bioinformatics tools that allow users to mine crop characterisation data (DIVSEEK).

Compared to the botanic garden community, the crop community is highly centralised around the FAO and the 11 multilateral germplasm collections in the gene banks of the Consultative Group on International Agricultural Research (CGIAR). Likewise, the International Treaty facilitates access to material and data between national gene banks, multilateral gene banks and users. No such centralised, multilateral infrastructure exists for botanic gardens. Nevertheless, there are strong parallels with the policy, infrastructural and collections frameworks that exist in the botanic gardens community. A botanic garden-centred Global System for the conservation and management of plant diversity would aim to collect, conserve, characterise and cultivate samples from all of the world's rare and threatened plants as an insurance policy against their extinction in the wild and as a source of plant material for human innovation, adaptation and resilience.

This Global System for botanic gardens would build on the following components:

- A global policy framework (the Convention on Biological Diversity).
- A global action plan (the Global Strategy for Plant Conservation).
- A review process (the Global Partnership for Plant Conservation).
- A collections infrastructure comprising an international network of botanic gardens and their living collections.
- A global portal of plant collection data (PlantSearch).
- An array of data sources providing access to phenotypic and genotypic data enabling conservation and use of the collections for human development and well-being.
- A range of tools, resources and activities that aims to increase awareness and participation in plant conservation resulting in wide-reaching benefits for society.

Most of the policy, planning and review architecture already exists, as indicated above. In addition, Botanic Gardens Conservation International (BGCI), which aims to coordinate the system (Smith, 2016), sits at the centre of a network of more than 500 botanic gardens in over 100 countries around the world. This network already holds globally significant *ex situ* collections, covering at least a third of the total known plant diversity. The PlantSearch database managed by BGCI on behalf of the global community of botanic gardens currently includes 1.3 million accession names from 1,141 botanic gardens around the world.[1] A recent comparison by Smith (2016), of the content of PlantSearch and *The Plant List* (2013) indicates that those gardens manage at least 115,787 different species in their living collections, equivalent to 33 per cent of all the species listed in *The Plant List*. This is already a remarkable collective asset which has the potential to be of even greater value in the future. There are, of course, caveats as mentioned in Chapter 4, such as the fact that accession records are not always kept up to date or accurately named and that PlantSearch itself is not comprehensive, covering only about 40 per cent of all botanic gardens.

Beyond these living collections, the botanic garden sector has impressive infrastructures. Kew's Millennium Seed Bank, the Royal Botanic Garden, Sydney's Plant Bank and Kunming Institute of Botany's Gene Bank of Wild Species are the largest, most sophisticated seed banks in the world and efforts are being made to encourage more botanic gardens to maintain local seed banks. The sector is equally strong in glasshouse and horticulture infrastructures and more than adequately served with micropropagation facilities and molecular laboratories. The international botanic garden community's most comprehensive data source on garden facilities and foci is BGCI's GardenSearch,[2] a web-based register of the world's botanic gardens comprising information on 2,671 botanic gardens and arboreta in 135 countries.

At the heart of the botanic gardens network is, as noted throughout this book, the knowledge and expertise of botanic garden staff. BGCI's GardenSearch database indicates that the world's botanic gardens employ at least 60,000 people, comprising thousands of plant scientists and horticulturalists who possess unique knowledge right across the taxonomic spectrum. This expertise is also manifest in specialist networks covering plant conservation practice such as red listing, seed conservation, tissue culture and cryopreservation, and ecological restoration. Such networks also include consortia with expertise in the conservation of specific groups of plants, such as orchids, succulent species, oaks and conifers.

As mentioned previously, the most comprehensive register of collection names is BGCI's PlantSearch database. The immediate aim is that PlantSearch becomes a portal to individual accessions and their data held in specific botanic gardens and ultimately becomes a means by which gardens can exchange material for conservation purposes in much the same way that the zoo community uses its International Species Information System as a stud book approach to captive breeding (Conde et al., 2011). PlantSearch 2.0, which will trial this approach, will be launched in 2017. A further BGCI database under development is 'ThreatSearch' which aims to be the most comprehensive consolidated list of plant threat assessments in the world. This database comprises global, regional and national assessments, and currently includes over 180,000 records; it was launched in late 2016. Together, GardenSearch, PlantSearch and ThreatSearch will be able to provide the underpinning informatics system necessary for the collaborative efforts of botanic gardens and partner organisations to comprise a Global System.

Notwithstanding the impressive array of resources already present or under development within the global community of botanic gardens, substantial investment will be required to build a fully functioning Global System that can prevent species extinctions in perpetuity. Perhaps the most important thing botanic gardens need to do is to agree, as a professional community, that they are going to be robust and united in taking on the challenge of plant species extinctions. Only by presenting a united front, and showing that commitment, are botanic gardens going to convince policymakers and funders that they have a substantial role to play.

Second, the community needs to promote plant conservation action in botanic gardens. Too often this activity currently competes with the other functions of botanic gardens, particularly with the need to increase visitors and generate income. Plant conservation activities in botanic gardens can be substantial or small and may include plant conservation policy, practice or education. What is important is that all botanic gardens do *something* – preferably plant conservation action, and with local relevance.

Third, plant conservation work in botanic gardens needs to be better coordinated, and botanic gardens need to focus their efforts on the gaps. It is critically important that they tackle the rarest, most threatened and most challenging species, especially those located in the same country or region as the botanic garden. Although the sector maintains a third of known plant diversity in its living collections and seed banks, improvements need to be made in strategically conserving and managing rare and threatened species. In a recent BGCI analysis of threatened tree species held in botanic gardens (Rivers *et al*., 2015) it was shown that only one in four threatened trees are present in *ex situ* collections. As well as covering the species gaps, botanic gardens also need to address the resource gaps by directing knowledge and financial resources towards smaller, resource-poor botanic gardens in biodiversity hotspots.

Fourth, the botanic garden sector needs to fully embrace the fact that it cannot succeed in maximising its potential in the era of global change by working in isolation. An *ex situ* seed or living collection is the means to an end, not the end in itself. The aim is to achieve self-perpetuating populations of plants within the broader landscape. This means working in an integrated way with other *in situ* conservationists (e.g. park managers, NGOs, etc.), foresters, farmers and other sectors that manage transformed landscapes. Explicitly, this also means that botanic gardens need to go out beyond their garden walls, and learn new disciplines. A large number of botanic gardens already manage wild areas and native species assemblages so this would not represent a huge ideological leap. However, botanic gardens have not always been as good at working in partnership with other professional sectors.

Finally, botanic gardens need to facilitate plant conservation action in broader society through stimulating public dialogue, creating opportunities for participation in local and global conservation efforts and provision of education, tools and information. At the same time, botanic gardens need to be careful that their plant conservation effort does not begin and end with public outreach and engagement. Currently too many gardens argue that they are fulfilling their role by simply informing the public about the need for plant conservation. This approach conveniently ignores the fact that the botanic garden sector has the technical skills that broader society does not, and that with those skills comes responsibility.

Standing back from the specifics of botanic gardens it easy to see that the establishment of a Global System is just one of many step changes that will be required to underpin the successful delivery of the 2030 Sustainable Development Goals, the internationally agreed path towards the Future We Want. Now that we have the ability to see how interconnected the best interests of humanity and the biosphere are, we hope that new understanding and new commitment will enable botanic gardens can take their place as key agents for undoing much of the damage we have inflicted on our planet.

Notes

1. See www.bgci.org/plant_search.php.
2. See www.bgci.org/garden_search.php.

References

Aronson, J. On Behalf of The ERA of Botanic Gardens. (2014). The ecological restoration alliance of botanic gardens: a new initiative takes root. *Restoration Ecology*, 22: 713–715.

Barham, E., Sharrock, S., Lane, C. and Baker, R. (2015). An International Plant Sentinel Network. *Sibbaldia*, 13: 83–97.

Blackmore, S. (2017). The Future Role of Botanic Gardens. In: Friis, I. and Balslev, H. (Eds), *Collections of Tropical Plants: Legacies from the Past or Essential Tools for the Future?* In press.

Corlett, R. T. (2016). A bigger toolbox: biotechnology in biodiversity conservation. *Trends in Biotechnology*, 35(1): 55–65, doi: http://dx.doi.org/10.1016/j.tibtech.2016.06.009

Conde, D. A., Flesness, N., Colchero, F., Jones, O. R. and Scheuerlein, A. (2011). An emerging role of zoos to conserve biodiversity. *Science*, 331, 1390–1391.

Elmqvist, T., Fragkias, M., Goodness, J., Güneralp, B., Marcotullio, P. J., McDonald, R. I., Parnell, S., Schewenius, M., Sendstad, M., Seto, K. C. & Wilkinson, C. (2013). *Urbanization, Biodiversity and Ecosystem Services: Challenges and*

Opportunities. Dordrecht, The Netherlands: Springer, 743 pp. doi: 10.1007/978-94-007-7088-1.

Food and Agriculture Organization (FAO). (2011) The Second Global Plan of Action for Plant Genetic Resources for Food and Agriculture. Commission for Genetic Resources for Food and Agriculture. Available online at: www.fao.org/agriculture/crops/thematic-sitemap/theme/seeds-pgr/gpa/en/ [accessed March 2017].

GPPC (2016). Progress in implementation of the targets of the Global Strategy for Plant Conservation 2014–2016. A report by the Global Partnership for Plant Conservation, UNEP/CBD.

Kapos, V., Balmford, A., Aveling, R. et al. (2008). Calibrating conservation: new tools for measuring success. *Conservation Letters*, 1: 155–164.

Kapos, V., Balmford, A., Aveling, R. et al. (2009). Outcomes, not implementation, predict conservation success. *Oryx*, 43(3): 336–342.

Pimm, S. L. and Joppa, L. N. (2015). How many plant species are there, where are they, and at what rate are they going extinct? *Annals of the Missouri Botanical Garden*, 100: 170–176.

Pinto, M., Almeida, C., Pereira, A. M., Silva, M. (2016). Urban Forest Governance: FUTURE: The 100,000 Trees Project in the Porto Metropolitan Area. In: Castro, P., Azeiteiro, U. M., Bacelar-Nicolau, P., Filho, W. L. and Azul, A. M. (Eds), *Biodiversity and Education for Sustainable Development*. Basel, Switzerland: Springer International Publishing, pp. 187–202.

RBG Kew (2016). *The State of the World's Plants Report 2016*. Richmond, UK: Royal Botanic Gardens, Kew, available online at https://stateoftheworldsplants.com/areas-important-for-plants [accessed February 2017].

Rivers, M., Shaw, K., Beech, E. and Jones, M. (2015). *Conserving the World's Most Threatened Trees. A Global Survey of Ex Situ Collections*. Richmond, UK: Botanic Gardens Conservation International.

Sharrock, S., Oldfield, S. and Wilson, O. (2014). *Plant Conservation Report 2014: A Review of Progress in Implementation of the Global Strategy for Plant Conservation 2011–2020*. Secretariat of the Convention on Biological Diversity, Montréal, Canada and Botanic Gardens Conservation International, Richmond, UK. Technical Series No. 81, 56 pp.

Smith, P. (2016). Building a Global System for the conservation of all plant diversity: a vision for botanic gardens and Botanic Garden Conservation International. *Sibbaldia*, 14: 5–13.

United Nations Development Programme (2015). *Transforming our World: The 2030 Agenda for Sustainable Development*. New York: UN.

Wandersee, J. H. and Schussler, E. E. (1999). Preventing plant blindness. *The American Biology Teacher*, 61, 82–86.

Index

access and benefit sharing 124, 126, 240
agarwood 53, 54
agroforestry 176
agronomic traits 169
Aichi targets 5, 25, 69, 134
Aldo Leopold 3, 13, 139
algae 118, 184–185
Anthropocene 1–3, 74–77, 88, 95, 96
apprenticeship 202
Aquilaria 53, 54
arbuscular mycorrhizal symbiosis 173
Aristoideae 171
Artocarpus altilis 176
ash dieback 106
assisted migration 89, 152
Atlanta Botanical Garden 137
audience 192–218
Australia 142
authentic engagement ix, 194–196

Banksia 143
basic human needs 193, 195
Betulaceae 63
bioclimatic envelope 127
biocontrol 138
biodiesel 180–182
biodiversity 73
biodiversity crisis 19
biodiversity hotspots 49
bioethanol 179–180
Biofuel Collections 182
biofuels 168, 178
Botanic Gardens Conservation International 9, 80, 244–245
breadfruit 176
Brundtland Report 4, 6
bryophytes 26, 62, 117
Bureau of Land Management 150

C_3 photosynthesis 170
C_4 photosynthesis 172
cacti 52, 62
carbon-fixation 170
Center for Plant Conservation 105, 135
centres of diversity 169
cereal production 167
cereals 170, 172
Chicago Botanic Garden 137
China 85
Chinese Academy of Sciences 95
Chinese Union of Botanic Gardens 95
Cincinnati Zoo & Botanical Garden 119
Cirsium pitcheri 137
CITES 4, 52–54
cities 240
citizen science 57, 127, 241
climate change 56–58, 78, 106, 127, 152–153, 168
closest living relatives 169
coevolution 36
community and landscape restoration 139–146, 240
community outreach 193
community restoration 139
conifers 62
conservation assessment 49–69
conservation impact 219–233
Convention on Biological Diversity 4
convergent evolution 170
crop wild relatives 168, 174–177
crops 106, 167
cryopreservation 113, 114
cryostorage 120
cryptic species 24, 25, 30
cultural institutions 195, 220
Cycads 62, 63

Dalbergia 16, 31, 53, 54
Darwin Initiative 106

Denver Botanic Gardens 127
desertification 126, 168
designer ecosystems 15
dicotyledons 62
Dinghu Shan 81, 82
DNA barcoding 26, 29–33, 238
DNA sequences 42
documentation 103
domesticated plants 106
dormant buds 115
Dutch elm disease 54

ecological restoration 18, 88–89, 126, 134–159, 240
Ecological Restoration Alliance 94, 157, 241
ecosystem function 69
ecosystem services 149
ecotypic adaptation 29
edge-effect 51
eDNA barcoding 33
education 185–187, 192–218
emerald ash borer 54
Encephalartos woodii 105
Endangered Species Act 59
endosymbiosis 172
engagement 192–218
environmental DNA 33
Erica verticillata 105
ethnobotany 89–90, 107
Euphorbia 31
European Consortium of Botanic Gardens 126
evaluation 219–233
ex situ conservation 17, 73, 128, 135
Exceptional Plant Species Advisory Group 120
exceptional species viii, 110, 116, 240
extinct 59
extinction 17, 50

Fabales 172
FAO 6
ferns 62, 117
fire regime 146
Flaveria 172
Flemingia stricta 142
floriculture 215–217
food security 126, 166–177, 201
forest monitoring 86–87
fossil fuels 178

fragmentation 51, 75, 79, 85
Francisco Javier Clavijero Botanic Garden 140
Franklinia alatamaha 105
Fraxinus 106
fuel security 177–185
functional trait diversity 147

gametophytes 118
Garden Search database 103, 245
Gene Bank of Wild Species 103
gene banks 176
genetic diversity 79, 104, 114, 117, 121, 147
genetic engineering 169, 173
genetic erosion 176
genetic resources 175
genetics 23–42
genomics 35
Gilbertiodendron maximum 83
Ginkgo biloba 31
global challenges 166–188, 200, 209, 219, 238
global change 2, 166–185
Global Crop Diversity Trust 109
global fuel security 166, 177–185, 241
Global Plan of Action 6
Global Strategy for Plant Conservation xiii, 9–12, 25, 68, 79, 83, 103, 111, 134, 221, 224, 230–233
Global System for Plant Conservation 242–246, 247
Global Tree Assessment 64
Gonostylus 31
Gran Canaria Declaration 10
'great acceleration' 74
Green Revolution 167
gymnosperms 62

habitat loss 51, 239
Hawaii 80
herbaria 23, 102
herbarium 30, 79, 84
Herbertus 26, 30
Herbertus borealis 30
herbivore 33, 36
high-throughput sequencing 26
Hong Kong 141
hongmu 53
horticulture 106
human diversity 242

hybridisation 25
Hymenoscyphus fraxineus 106

important plant areas 84
illegal harvesting 24
illegal logging 53
illegal trade 239
impact assessment 233
Indonesia 106
influencing policy 68
Inga edulis 26
Inga 36
in situ conservation 17, 73–98, 221
International Centre for Agricultural Research in Dry Areas 109
International Network for Seed Based Restoration 150
International Plant Exchange Network 126
International Plant Sentinel Network 55, 106, 127
International Plant Treaty 5
introgression 25
invasive species 34, 51, 55, 81, 106, 142, 146, 153
inventories viii, 82
in vitro propagation 113
IPBES 5
IUCN 3
IUCN protected area categories 76
IUCN Red list 65

Jatropha curcas 183–184
Jerusalem Botanic Gardens 205–209
John Muir 3, 4

Karrikins 148
Kenya 106
Kings Park and Botanic Garden 119, 142, 143–146

landraces 169
legumes 173
Limbe Botanic Garden 215–217
living collections 102, 168
local communities 91

Madagascar 53, 83, 93–94
Magnoliaceae 62
Malawi 106
Malaysia 83
Man and the Biosphere 12

Material Transfer Agreements. 126
measures of diversity 147
medicinal plants 16
metabarcoding 31
Mexico 140
Michelia maudiae 142
middle and high school 211
migration rates 16
Millennium Development Goals 7
Millennium Seed Bank 108
Mimulus 36
Mimulus guttatus 38
Minuartia cumberlandensis 119
Missouri Botanical Garden 83
monitoring 85–88, 151–152, 241
monocotyledons 62
Morton Arboretum 91, 137
Mulanje Cedar 106
Multilateral Environmental Agreements 4

National Biodiversity Strategies and Action Plans 68
National Phenology Network 127
National Seed Strategy 150
National Tropical Botanical Garden 80, 107, 176
'native winners' 146
natural resources 89–90
nature-based solutions 201
networking viii, 94
New York Botanical Garden 118
New Zealand 106
nitrogen fixation 172
non-timber forest products 52
North American Strategy for Plant Conservation 222
novel ecosystems 14

oak processionary moth 106
orchids 52, 137
orthodox seeds 107
outreach 185–187
overexploitation 54, 239

Pedicularis 36
perceptions 223
performance measures 224
phenological diversity 148
phylogenetic analysis 171
phylogenetic diversity 148
planetary boundaries 2

Plant Bank 103
plant blindness 241
plant breeding 188
Plant Conservation Alliance 150
plant diversity 219
Plant List 25, 238
Plant Treaty 124
plant–animal interactions 33
PlantSearch database 102, 111, 126, 245
Platanthera integrilabia 137
pollen 32, 102
polyploidy 29, 36–38
population 167, 177, 239
Potentilla robbinsiana 137
precautionary principle 13
prior informed consent 125
Project BudBurst 57
protected areas 14, 75, 78
provenance 121, 151
Pterocarpus 16, 53, 54
public dialogue 246
public engagement 185–187
public perception 223

Quercus 105

rare plant reintroductions 135–139
'recalcitrant' seeds 108
Red List 3
red listing 17, 24, 50, 58, 83, 239
reintroduction 104
resilience 148
restoration 128
'restoration ready' seed 149
Rhamnus cathartica 158
rice 167
root nodulation 172–173
root nodulation symbiosis 172
Rosales 172
rosewood 16, 53
Royal Botanic Garden Jordan 154
Royal Botanic Garden Edinburgh 117
Royal Botanic Gardens, Kew 94, 105
Royal Botanical Gardens Hamilton and Burlington 140
RuBisCO 170

Sample Red List Index 62
seed banks 17, 73, 102, 106–116, 176
seed viability 108

Seeds of Success 150
Senecio lautus 39
simple sequence repeats 25
Singapore 83
slash and burn 51
social inclusion 198
social justice 201–205
somatic embryos 116
South Africa 84
South African National Biodiversity Institute 83, 84, 152
South China Botanic Garden 81, 142
speciation 25, 35
species discovery 23
species estimates 23
species identification 23–29, 40
species management 228
species recovery programmes 240
spore banking 117
Strategic Plan for Biodiversity 12
subtropical forest 141
Sustainable Development Goals 7, 233, 236–238, 247
sustainable farming 201
sustainable lifestyle 205
Svalbard Seed Vault 109
Symplocos coccinea 141

tallgrass prairie 148
taxonomy and systematics 13, 23–29, 36, 80, 83, 134, 172, 224
The Future We Want 7
The Russian Federation of Botanic Gardens 104
threat assessment 49–69
threat categories 60
Threat Search database 245
tissue culture 114
translocation 104
tropics 73

UNEP 4
UNFCCC 4, 7
United States Geological Service 137
University of Washington Botanic Garden 152
University of Wisconsin Arboretum 139
urban environment 15, 77
urban forest 91

urban forests 240
urban resilience 206

Verband Botanischer Gärten 126

Widdringtonia whytei 106
wildlife crime 24, 31, 53, 239
Windy City Harvest 201–205

Woodsia ilvensis 137
World Flora Online 25, 238
WWF 3

Xishuangbanna Tropical Botanic Garden 80, 92–93, 241

'Zero Extinction Project' 95